W9-AVR-834

MARS
ON
EARTH

ALSO BY ROBERT ZUBRIN

Entering Space: Creating a Spacefaring Civilization

The Case for Mars: The Plan to Settle the Red Planet and Why We Must (with Richard Wagner)

Islands in the Sky: Bold New Ideas for Colonizing Space (with Stanley Schmidt)

First Landing (a novel)

ROBERT ZUBRIN

Jeremy P. Tarcher/Penguin
a member of Penguin Group (USA) Inc.
New York

MARS ON EARTH

The Adventures of Space Pioneers in the High Arctic

Most Tarcher/Penguin books are available at quantity discounts for bulk purchase for sales promotions, premiums, fund-raising, and educational needs. Special books or book excerpts also can be created to fit specific needs. For details, write Penguin Group (USA) Inc. Special Markets, 375 Hudson Street, New York, NY 10014.

Jeremy P. Tarcher/Penguin
a member of
Penguin Group (USA) Inc.
375 Hudson Street
New York, NY 10014
www.penguin.com

Published simultaneously in Canada

Library of Congress Cataloging-in-Publication Data

Zubrin, Robert.
Mars on Earth : the adventures of space pioneers in the high
Arctic / Robert Zubrin.
p. cm.
Includes bibliographical references and index.
ISBN 1-58542-255-X
1. Space flight to Mars—Research—United States. 2. Research
institutes—Arctic Regions. 3. Mars—Exploration—
Planning. 4. Devon Island (Nunavut)—Description and travel.
5. Discoveries in geography. 6. Mars Society. I. Title.
TL799.M3Z83 2003 2003050757
629.45'53'072—dc21

Printed in the United States of America
10 9 8 7 6 5 4 3 2 1

This book is printed on acid-free paper. ♾

BOOK DESIGN BY MEIGHAN CAVANAUGH

To my friends,

the Mars Society Volunteers—

Soldiers of Hope

Dreamers and Seekers

Fighters for the Future

Then felt I like some watcher of the skies
When a new planet swims into his ken;
Or like stout Cortez, when with eagle eyes
He stared at the Pacific—and all his men
Looked at each other with wild surmise—
Silent, upon a peak in Darien.

—JOHN KEATS, *"On First Looking into Chapman's Homer"*

CONTENTS

ACKNOWLEDGMENTS

The author wishes to thank Chris McKay, Imre Friedmann, Bill Clancey, Judith Lapierre, and Ethan Vishniac for their critical review of parts of the manuscript that led to the present work. I also express my thanks to my editor, Mitch Horowitz, whose guidance in the critical task of transforming a huge mass of raw manuscript material into a coherent book was invaluable. Thanks are also due to my agent Laurie Fox, without whose coaching and brilliant salesmanship this work might never have been published.

A book about an adventure would not be possible without the adventure itself. Space limitations preclude naming all of those who contributed to the Mars station project, but special mention of some is appropriate here. These include Pascal Lee, who first suggested that the Mars Society initiate such a program; Kurt Micheels, whose architectural work formed the basis of the Flashline Arctic Station (and to a lesser extent Desert Station) design; John Kunz and Scott Fisher, who respectively developed the technology used for the construction of the Flashline and Desert stations; the United States Marines for performing the difficult Arctic paradrops needed to bring the Flashline Station components to Devon Island; Aziz Kheraj and the people of Resolute Bay for providing logistical support for the Flashline Station; Don and Connie Foutz and Larry and Lamont Ekker for providing logistical support for the Desert Station; Frank Schubert and Matt Smola for leading the construction of both stations; Andy Liebmann of Resolute Films; Steve Burns of the Discovery Channel; Charles and Louise

Stack of the Flashline Corporation; Steve and Michelle Kirsch; Rick Tumlinson of the FINDS foundation; the United Association of Plumbers and Pipefitters and the International Sheetmetal Workers Union; Adrian Berry Lord Camrose; Elon Musk; Eric Tilenius; James Cameron; Buzz Aldrin; Paul Young and Starchaser Industries; Jim Greenleaf of the Greenleaf Machine Tool Corporation; and the Kawasaki, Celestron, Bushnell, Alcoa, and Generac corporations, all of whom were essential in obtaining or providing the funds and equipment needed to build and operate the stations.

I wish to give special thanks to Joe Amarualik and all the Inuit of Resolute Bay who joined with Mars Society volunteers to build the Flashline Station, and to Anna Paulson and her group of Mars Society volunteers, who helped to build the desert station. Additional thanks are due to Peter Detterline, who led the effort to create the Mars Society Desert Observatory; Gary Fisher, who spearheaded the development of the Greenhab bioregnerative life support module at the Mars Desert Research Station (MDRS); and Pat and Tom Czarnik, who led and coordinated the group of Mars Society volunteers who ran the MDRS while it was on exhibit at Kennedy Space Center. Special thanks are also due to the members of the Rocky Mountain Mars Society (RMMS), who mobilized to complete the Flashline Station manufacture in time for ship-out, developed the spacesuit simulators used at our stations, and formed the central core of the Mars Society Mission Support team. RMMS members whose names cry out for particular mention include Tony Muscatello, the Director of Mars Society Mission Support, and volunteers-without-equal Lorraine Bell, Brian Enke, Robert Pohl, Patty Piteau, Dewey Anderson, Gary Snyder, Emily Burrows, Jason Held, and Paul Graham. Other Mars Society chapters who have shared the work of Mission Support include the San Diego Chapter, the Northern California Chapter, and the Toronto Chapter.

As of this writing, some 140 volunteers have served as crew members of either the Flashline Arctic or Desert Station, and their experiences and observations form a major contribution to this narrative. I thank them all. I would, however, be remiss if I did not take this opportunity to express my heartfelt appreciation to my own crewmates, whom I had the honor to command in Flashline Station rotations Two, Three, and Seven, and Mars Desert Research Station (MDRS) rotation One. These include: in Flashline Two, Steve Braham, Vladimir Pletser, Bill Clancey, Katy Quinn, and Charles Cockell; in Flashline Three, John Blitch, Cathrine Frandsen, Christine Jayarajah, Brent Bos, and Charles Frankel; in MDRS One, Steve McDaniel, Heather Chluda, Jennifer Heldmann, Frank Schubert, and Troy Wegman; and in Flashline Seven, Nell Beedle, Markus Landgraf, Shannon Hinsa, Emily MacDonald, K. Mark Caviezel, and Frank Eckardt. Dear crewmates: thanks for your dedication, hard work, and companionship, and for putting up with an ornery character like me.

Finally, I wish to express my deepest thanks to the members of my family—daughters Rachel and Sarah, and dear wife Maggie, whose love and affection sustained me through the years of effort reported herein. Maggie's role, however, went well beyond that of a supportive wife. In addition to holding our family together, she did more than any other single person to hold the Mars Society together through many difficult and fractious times. Without her, none of this would have been possible.

1. THE CHALLENGE OF MARS

Our doubts are traitors,
And make us lose the good we oft might win
By fearing to attempt.

—SHAKESPEARE, *Measure for Measure*

The Earth is not the only world. There are billions of other potential homes for life. The first of these is now within reach.

The planet Mars is a world of towering mountains, vast deserts, polar ice fields, dry river channels, and spectacular deep canyons. Possessing a surface area equal to all the continents of the Earth put together, it orbits our sun at a distance about 50 percent greater than that of the Earth. This makes Mars a cold world, but not impossibly so. The average sunlight received at the Martian equator is about equal to that which shines upon Norway or Alaska. During the day at low Martian latitudes, the temperature frequently exceeds 50°F (10°C). At night, however, the thin Martian atmosphere does a poor job of retaining heat, and temperatures drop to –130°F (–90°C).

There is no liquid water on the surface of Mars today, but there was once, and our satellite probes show us its handiwork in the form of large networks of dried-up riverbeds, dry lakes, and even the basin of a now-vacant northern Martian ocean. The water, however, is there, its surface reserves frozen as ice and permafrost and covered with dust, its deeper reservoirs still liquid, warmed by the planet's remaining sources of geothermal heat. There is as much water per square mile on Mars as there is on the continents of our home world.

Water is the staff of life, and the presence of large quantities of water on Mars mark it out as a potential home for a biosphere. On Earth, wherever we find liquid water, we find life. The evidence from our orbital images shows that there was liquid water on the surface of Mars for about a billion years of the planet's early history, a span roughly ten times as long as it took for life to appear in the Earth's fossil record after there was liquid water here. Thus if the conjecture is correct that life is a natural development from chemistry wherever one has liquid water and a sufficient period of time, then life should have appeared on Mars. Fossils recording its history may be findable.

Life may have lost its foothold on the planet's surface when the climate deteriorated with the loss of the juvenile Mars's early thick carbon dioxide atmosphere and its associated greenhouse warming capability. But our space probes show that liquid water has gushed out from the Red Planet's subsurface within the past few million years—effectively, the geologic present. This means that refuges for retreating Martian life still exist. If we go there and drill, we can find them, and in finding them determine whether life as we know it on Earth is the pattern for all life everywhere or whether we are just one example of a much vaster and more varied tapestry.

Mars is thus the Rosetta Stone that will reveal to us the nature of life and its place within the cosmic order.

THE NEW WORLD

But Mars is more than just an object of scientific inquiry. It is a world capable of sustaining not only an ancient native microbial ecology, but a new immigrant branch of human civilization. For the Red Planet's resources go well beyond its possession of water. It has carbon in abundance as well, present both in the carbon dioxide that composes the majority of its atmosphere and in carbonates in its surface material. It has nitrogen too; nitrogen is the leading minority gas in Mars's air and almost certainly exists as nitrates in the soil as well. Thus between the water, carbon dioxide, and nitrogen, we have all four of the primary elements of life (carbon, nitrogen, oxygen, and hydrogen). Calcium, phosphorus, and sulfur, the key secondary elements of life, are also present in abundance. (In contrast, with the exception of oxygen bound as oxides in rock, all of these are either rare or virtually absent on the Earth's Moon.)

In addition, all the elements of industry—such as iron, titanium, nickel, zinc, silicon, aluminum, and copper—are available on Mars, and the planet has had a complex geological history involving volcanism and hydrological action that has allowed for the concentration of geochemical rare elements into usable concentrated mineral ore. The Mars day-night cycle is 24.6 hours long, nearly the same as the Earth's, which is not only pleasant for humans, but more importantly, makes it fully suitable for growing plants in outdoor greenhouses using natural sunlight. The planet's geothermal heat, which currently may sustain the habitats for scientifically fascinating native microbes, can also be used to provide both plentiful liquid water and power for human Mars settlements.

In a way that simply is not true of the Earth's Moon, the asteroids, or any other extraterrestrial destination in our solar system, Mars is the New World. If we can go there and develop the craft that allows us to transform its native resources into usable materials—if we can transform its carbon dioxide and water into fuel and oxygen, if we can use its water and soil and sunlight to grow plants, if we can extract geothermal power from its subsurface, if we can use its collection of solid resources to produce bricks, ceramics, glasses, plastics, and metals, and then move up the ladder of craftsmanship to produce these things in complex shapes to make wires, tubes, clothes, tankage, habitats, and so forth—then we can create the technological underpinnings for not only a new branch but a new *type* of human society.

Because it is the closest world that can support settlement, Mars poses a critical test for the human race. How well we handle it will determine whether we remain a single planet species or become spacefarers with the whole universe open before us.

THE ROAD NOT TAKEN

> We conclude that NASA has the demonstrated organizational competence and technology base, by virtue of the Apollo success and other achievements, to carry out a successful program to land man on Mars within 15 years. There are a number of precursor activities necessary before such a mission can be attempted. These activities can proceed without development specific to a manned Mars mission—but for optimum benefit should be carried out with the Mars mission in mind. We conclude that a manned Mars mission should be accepted as a long-range goal for the space program. . . . NASA has outlined plans that would include a manned Mars mission in 1981. . . . Such a program would result in maximum stimulation of our technology and creation of new capability.
>
> —Report of the President's Space Task Group, September, 1969

We should have been on Mars already.

In February 1969, with the Apollo space program nearing its Moon-landing climax, newly elected President Richard Nixon directed the formation of a presidential-level Space Task Group (STG) and charged it with the responsibility of drawing up the program for the continued activities of NASA for the next two decades. The group—chaired by Vice President Spiro Agnew and consisting of Air Force Secretary Robert Seamans, visionary NASA Administrator Thomas O. Paine, Presidential Science Adviser Lee Dubridge, and AEC Chairman Glenn Seaborg—examined a series of in-depth studies prepared by NASA, the Department of Defense, the National Academy of Sciences, and the American Institute of Aeronautics and Astronautics (AIAA) before issuing its summary report in September 1969. The report was then presented to the public in detail in a long article published in the January 1970 issue of *Astronautics and Aeronautics* magazine. The program was remarkable. Looking back on it today, one can almost feel the melancholy that a ninth-century Italian might have felt in looking on the ruins of Imperial Rome.

Had this plan not been aborted, America's space program today would be yielding us the benefits of manned research stations in many locations in space, including large permanent bases on both the Moon and Mars. November 12, 1981—the day that in actuality the Space Shuttle was launched for the second time—was to have been the date when piloted craft were to have left the Earth-orbiting space station on the first human expedition to Mars.

The report projected the following:

Space Stations: The program called for continued production and use of the Saturn V moon rocket to launch in 1972 a space station eventually known as Skylab. The biomedical and engineering data from this outpost would provide the basis for the design of a larger Space Station module, which would be launched as a permanent manned orbital base by 1976, and then enlarged, through successive addition of modules, to form a hundred-person space base by 1980. Such a space base would serve as housing for scientists engaged in astronomical, biological, physical, chemical, and materials-processing research, as well as platforms for the launching and repair of satellites, and manned and unmanned lunar and interplanetary probes.

Space Shuttle: In 2002 dollars, a Saturn V could deliver payload to Earth orbit for about $7,000 per kilogram. (In this book I shall use kilograms and tonnes for units of mass. One kilogram equals 2.2 pounds. One tonne equals 1000 kilograms, or 2200 pounds.) The STG plan called for the rapid development of a fully reusable Space Shuttle by 1975 that could launch payloads to Earth orbit for $350 per kilogram. Whether this could have been achieved is debatable. It should be pointed out, however, that the STG's Space Shuttle was envisaged as part of an overall space initiative that would have provided the Shuttle fleet with scores of launches per year, enough to make its feature of full reusability a real cost saver. (Reusable launch vehicles with low launch rates produce no savings, and in fact are more expensive than expendables because their large ground support force must be paid the same regardless of how many launches occur.) The Shuttle program was initiated. Because of budget cuts, however, the program was stretched out and the vehicle made only partially reusable, and many other corners were cut. As a result of these technological compromises and the low launch rate for the Shuttles provided by the truncated NASA program overall, the actual launch cost achieved by the Shuttle fleet is about $20,000 per kilogram.

Nuclear Rocket Engine: The NERVA nuclear thermal rocket engine had demonstrated 62 minutes of continuous firing in ground tests in 1968, and was scheduled to be fully qualified for use in space operations by 1978. The NERVA's exhaust velocity was about double that of the best possible chemical engines (9 kilometers per second compared with 4.5 kilometers per second), allowing it to perform orbit transfers using half the amount of propellant.

NERVA stages would have been permanently stationed in space, refueling at the Space Station from propellant stocks delivered there by Shuttle or Saturn V. The NERVA vehicle would haul space station modules to be permanently stationed in geosynchronous orbit by 1980, and would act as an ongoing orbit transfer capability—hauling propellant, people, and materials from the Earth orbital to the Moon orbital stations. In 1981, three NERVA engines were to be used to power a six-man spacecraft to Mars and back.

Space Tug: The space tug, a small vehicle powered by a chemical rocket, was scheduled to be completed by 1975. Until the NERVA stage became available, it could also be used as an orbit transfer vehicle, and would be used to move a six-man space station module out to lunar orbit as early as 1976. The space tug would also be capable of landing and taking off from the surface of the Moon, and thus act as a shuttle from the

Moon base to the lunar orbit station. It would also be able to set a space station module down on the surface of the Moon to be used as a base.

Lunar Exploration: Apollo manned lunar expeditions would continue through the 1970s, with improved equipment being used, such as pressurized ground rovers to lengthen the stay and expand the areas of exploration. The development of the NERVA, the space tug, and the lunar orbit station would make possible a permanent six-man Moon base by 1979, expanding to forty-eight people by 1983, at which point the lunar orbit station would have a crew of twenty-four. The inhabitants of the Moon base would engage in all kinds of astronomical and geological research, as well as develop techniques for mining and processing lunar materials. Because of the Moon's low gravity, it was anticipated that any capability for processing lunar material for use in space would significantly reduce the cost of the entire space enterprise.

Mars Exploration: November 1981 was chosen as the target date when two six-man spacecraft would leave the low Earth-orbit space station on a trans-Mars trajectory. The Mars ships would consist of three standard NERVA stages attached to a space station module, with a specialized Mars Excursion Module (MEM) attached. Two of the NERVA stages would almost exhaust themselves to launch the ship on its trajectory toward Mars. Then they would detach and fly back under automatic guidance to Earth orbit, where they could be replenished for another use. Meanwhile the ships would coast on out to Mars, arriving during August 1982, to be braked into Mars orbit by the remaining NERVA stage. Each ship would then launch six sterile automatic probes to various parts of the Martian surface, which would land, collect soil samples, and return to the ships for analysis.

After thus being assured of the lack of any biological hazard, the Mars excursion modules would land with three men from each of the crews for a forty-day stint of exploration on the Martian surface. After that period, the Mars excursion modules would then return to the mother ships in orbit and the remaining NERVA stage would then power the escape from

Mars orbit, aiming the ships on a 300-day trajectory back to Earth, where they would park in orbit and the astronauts would debark.

Further such expeditions were scheduled for 1983 and 1986, with a portion of the 1986 crew staying behind to establish a permanent Mars base. By 1989 the United States intended to have a permanent Mars base crewed by forty-eight people.

But that was the road not taken. To implement it would have required continuing the space program at Apollo levels of funding for the twenty years following 1969. Instead, the Nixon administration aborted the program. The Mars mission, lunar base, NERVA program, and space tug were all canceled. The Space Shuttle program was degraded. The Saturn V assembly line was shut down, its tooling was destroyed, and the blueprints were given to a Boy Scout paper drive in Huntsville, Alabama. The funds that could have opened the solar system to humanity were thus "saved," and used instead to provide a fraction of the resources needed to allow the South Vietnamese government to linger on for a few more years.

By the early 1980s, the idea of a pioneering space program was a fading memory. I can remember a scientific meeting I attended in New York City during that time, which was addressed by a politically appointed senior NASA official from Washington, D.C. The purpose of NASA, he said, was to use the Space Shuttle to launch communication, weather, and military reconnaissance satellites.

I asked, "But what of the vision of humans exploring the Moon and Mars?"

He replied that those were the dreams of youth—we were wiser and more mature now.

At this point he was blasted by a female astronaut who was in attendance. No, she said passionately, the purpose of NASA should be to open the space frontier, and we—those in NASA who still stand for that vision—need you, the scientists, the engineers, and the public who believe in it, to fight politically to get that kind of space program.

It was good the astronaut got her licks in then. Her name was Judith Resnik. In January 1986 she was killed aboard the *Challenger*, a victim of the penny-pinching design compromises forced upon the Space Shuttle program by the Washington, D.C., space policy set.

THE CONSEQUENCES OF RETREAT

The decision of the Nixon administration to allow the American space program to collapse after the Moon landing represented a catastrophic failure of political leadership. We had the solar system in our grasp, and we threw it away. It was as if after Christopher Columbus had returned from his discovery of the New World, Ferdinand and Isabella responded by saying, "Let's burn the fleet." While over the following years certain factions within the Reagan and first Bush administrations attempted to rally the space program for a second try, so far none have been able to rebuild the house that Jack Kennedy built, and the dominant political forces in the various administrations and congresses have effectively and repeatedly reratified Nixon's failure.

As a result, during the thirty years after Apollo sent men 400,000 kilometers from the Earth, far from achieving the STG plan's thousandfold increase in exploratory range necessary to reach Mars, no human has voyaged out more than 400 kilometers. We have been going nowhere.

It has not been a question of money. The Reagan and first Bush administrations were friendly to NASA, and raised its budget to levels exceeding, in real inflation-adjusted dollars, the average NASA budget of the Apollo years. The Clinton administration reduced NASA's budget only slightly below those levels. In today's money, the average NASA budget over the period 1961 to 1973 (from JFK's speech kicking off the Apollo program to the final Skylab missions) was about $16 billion. NASA's budget this year will be about $15 billion. So the expenditures the political establishment has been willing to provide are completely comparable to what was actually spent to send people to the Moon, and projected as necessary to realize the STG plan afterward. What has been lacking is leadership—or more specifically, courageous leadership.

It takes courageous leadership to set a goal, because if you set a goal, you must also accept the risk that you may fail to reach it. But there can be no progress without one. In the 1960s, NASA had a goal that focused all its technology development efforts and forced them to produce real hardware that could be used effectively in combination to achieve real results. In consequence, in the course of going to the Moon, we also developed hydrogen oxygen rocket engines, multistaged heavy-lift launch vehicles, in-space life-support systems, spacesuits, space navigation and communication systems, orbital rendezvous and docking techniques, space telescopes, robotic probes, space nuclear power and propulsion systems, lunar landing systems, lunar rovers, reentry systems—in fact, the entire array of space technologies we use today. We also flew some forty lunar and planetary probes, created several major space research centers, and inspired an entire generation of youth to enter science and engineering.

In the 1990s, without a real aim, NASA spent as much money as it did in the 1960s, but it did not send people beyond low Earth orbit, it developed no new technologies to speak of, it flew only about ten planetary probes, it created no important new centers of research, and it hardly inspired anyone.

To see how the two different approaches to conducting a space program play out, consider the case of launch vehicle technology.

During the 1960s, NASA needed to go to the Moon. So they designed a set of spacecraft, estimated their weight, and on that basis drew up specifications for a heavy-lift booster, the Saturn V, capable of launching them. Since the Apollo program had a deadline to land astronauts on the Moon by the end of the decade, and several preliminary flights were needed before the actual Moon shot, they put the Saturn V program on a schedule to be operational by 1967. So they started the program in 1962, and moved it on track to achieve first flight in 1966.

In the 1990s, NASA started a program called X-33 "to develop revolutionary new launch vehicle technology." The program was started in 1996, and canceled in 2001 without achieving significant results.

So four years working on Saturn V achieved an operational heavy-lift launch vehicle; five years working on X-33 achieved nothing. If the Saturn V program had failed, crisis would have resulted, as there would have been no means to send the astronauts to the Moon. But the failure of the X-33 program troubled no one, because it wasn't really needed for anything.

In the course of the Apollo 13 mission, Johnson Space Center Flight Director Gene Kranz declared famously, "Failure is not an option." In the rudderless modern NASA, it might be said that failure is not a problem.

In 2002, NASA Administrator Sean O'Keefe elevated the space agency's modern goal-free approach to spending the taxpayer's money to a sacred principle. "NASA," he said, "should not be destination-driven."

APOLLO'S CHILDREN

John Kennedy is gone, and the quality of leadership bred and tested in the crisis of the Depression and World War II is no longer available in Washington. Left to itself, it is thus questionable whether the political system will again produce a figure capable of setting the nation's sights on the stars.

But leadership can come from many places. It does not need exalted political office, or high military rank, or expensive pinstripe suits. All it needs is an idea whose time has come, and the courage to see it through.

There are those of us whose first memories are of Sputnik, or Kennedy's Moon speech, or John Glenn's flight. There are those of us who grew up in a time when anything seemed possible: we were going to the Moon by 1970, Mars by 1980, Saturn by 1990, Alpha Centauri by the year 2000—those of us who understood that the true purpose of the Apollo program was not to meet the petty political needs of the present but to open the prospects of an unlimited future, who saw the spirit moving the Apollo program and heard its call to challenge the infinite frontier. It is said that the Apollo program inspired millions of the youth of its day, and this is true in the most literal sense: Apollo put its spirit inside of them. Apollo's children saw their dreams betrayed by the politicians. But Apollo's children did not give up.

This book is about an effort to change history. It is about a group of people called the Mars Society, composed largely of Apollo's children and grandchildren, who are attempting with very limited resources to alter the course of human events in a rather substantial way. I am one of the leaders of this group, so much of this account will be in the first person. Put simply, our goal is to put the exploration and settlement of Mars on mankind's agenda.

Our approach toward accomplishing this revolution is unique. Rather than try to seize power, we have launched a science project. This program, which involves setting up stations in remote areas where we can learn how humans will need to live, work, and explore on Mars, has great scientific importance in itself, and will therefore be reported on in considerable detail. It is our hope and belief, however, that by using this activity to make sensuous the vision of human Mars exploration, we can act as a rallying point to mobilize the hundreds of millions of people across the globe who believe it is essential to a positive future that humanity expand into space. As we shall see, we are already beginning to have some success in this regard.

The Mars Society's project to discover how humans will live and work on Mars in the future has its origins in earlier work by others who used their studies of remote and hostile environments on Earth to try to learn how simpler life forms might be living on Mars today. We begin with their story.

2. SEARCHING FOR MARTIAN LIFE ON EARTH

Asgard Mountains, Antarctica, December 10, 1973

The missing person's name was Wolf Vishniac.

There was reason to be concerned, for Vishniac was a driven man.

Born of Jewish parents in Berlin in 1922, Vishniac had seen his world collapse before. In fact, his father, the famous photographer Roman Vishniac, had documented the debacle in books depicting the disintegration of Jewish life in Central Europe in the years preceding the war. His family had escaped to America, and Wolf had launched a career as an eminent microbiologist. But now, despite scientific talent of the highest order, his professional life was in shambles.

Vishniac was one of the founders of scientific exobiology. As early as the 1950s, he realized that no instrument existed that could detect life on other worlds, so he set about designing one. In 1959, he received NASA's first exobiology instrument grant, some $4,500 to start work on a life-detection device that eventually became known as the Wolf Trap. As powerful space instruments go, the Wolf Trap was a model of elegant simplicity. Basically, it was just a tube containing a transparent liquid culture medium. A tiny sample of Martian dirt would be dropped in, and a light shined through the fluid. The transparency of the fluid and its pH, or acidity, would both be measured. When bacteria multiply in a clear culture medium, they make it cloudy and change its pH. If either quality changed substantially over time after the liquid was contacted by Martian material, that would indicate the presence of life.

In 1963, the Wolf Trap was selected as one of the life-detection experiments to be flown on NASA's planned large unmanned Mars landing mission called Voyager (no relation to the outer solar system exploration probe of the same name that flew in the 1970s). Unfortunately, Voyager was canceled in 1967. A new smaller Mars lander project, called Viking, was started in 1968, however, and after another round of competition, Vishniac's Wolf Trap was selected for inclusion in this program in 1969.

One can only imagine the frustration Vishniac must have felt during the 1960s, as the years went by, attending or chairing one committee meeting or review after another, trying to keep the program intact, only to watch it collapse completely in 1967, and then be forced to compete his instrument for inclusion all over again in 1969. One can only imagine the stirrings of elation he must have started to feel as the seventies opened up with the program moving ahead toward its planned summer 1973 launch date.

It takes very little imagination, however, to guess how Vishniac must have felt in March 1972, when he was informed by NASA headquarters that his instrument was being pulled from the mission.

The decision was outrageous. The cover story was that the deletion of the Wolf Trap was necessary to save mass and cost on the Viking lander, both of which were ballooning out of control. But all the real expenditure involved in developing the Wolf Trap had already been made, and the instrument actually weighed very little. As Viking Science Team leader Harold "Chuck" Klein said at the time in a letter protesting the decision to NASA Headquarters' Associate Administrator for Space Science John Naugle: "This science reduction is all the more difficult to accept because it is not at all clear just what factors dictated this decision. Recent discussions with TRW . . . leave little doubt that no savings in weight or volume will follow from the elimination of the light scattering experiment [i.e., the Wolf Trap]. . . . Whether, at this late date, any cost savings will accrue from the deletion is also problematical."[1]

Naugle claimed that the elimination of the Wolf Trap would save $2.3 million. But the Viking mission cost $2 billion, and eliminating one of its four life-detection instruments in order to save one-thousandth of the program budget would appear to be a poor trade indeed. In my view, the real reason for the pulling of the Wolf Trap could only have been an effort on the part of NASA HQ to generate some ammo to impress upon Congress the serious damage their budget limits were doing to the program.

Be that as it may, thirteen years of Vishniac's work was down the tubes. Furthermore, his NASA funding disappeared, and to make matters worse, he discovered that the administrators of the National Institute of Health and the National Science Foundation nursed a grudge against those biologists who had "gone NASAing around" during the heady space-money days of the 1960s. Thus, as a result of his decade-long close association with NASA, Vishniac found it difficult to obtain alternative support from the more traditional biology research agencies that had previously financed his work. Vishniac expressed his desperation at that time to fellow Viking scientist Gerald Soffen: "It is essential that I recapture some sort of standing in the academic world and I must therefore limit my participation in Viking to essentials only."[2]

But the purpose of the Wolf Trap had not been to provide a meal ticket. Its purpose had been to prove the universality of life. If that could not be done by sending the instrument to Mars, perhaps it could be done another way, by sending it instead to the most Mars-like environment on Earth.

The dry valleys of Antarctica's South Victoria Land are Earth's most extreme cold-desert region, and as such are generally regarded as one of our planet's best twins for duplicating the environment of Mars. The valleys are several thousand square kilometers in extent, and are kept ice-free by the Transantarctic Mountains, which cut off the flow of glaciers from the continent's interior. The average mean annual temperature is about –25°C, with a peak air temperature just above 0°C during the dog days of summer. There is no rainfall, and practically no water vapor in the air. A tiny amount of snow does fall, but because of the lack of water vapor pressure in the atmosphere, it usually sublimes away without ever melting into liquid water. The ultraviolet flux, deadly to microorganisms, is also higher than most other places on Earth.

The first scientist to study the dry valleys with a view toward their significance regarding the life on Mars question was Roy Cameron of NASA's Jet Propulsion Lab (JPL). Cameron, a microbiologist with support from the National Science Foundation, took a team to Antarctica in the austral summer of 1965–66 to begin a three-year program of intense study of Southern Victoria Land. Basing himself at McMurdo Sound, Cameron used Weasel tractors, Sno-Cats, and helicopters to carry out a series of sorties to the dry valleys, some 160 kilometers away, and through his extended and intense effort, managed to collect 290 samples from 135 sites representing many different locations and soil depths. The samples were all collected with exacting care using only sterilized digging tools, and then, packed in containers that kept them at a representative Antarctic temperature of –25°C, shipped back to Pasadena, California, where they were analyzed by Cameron's collaborator Norm Horowitz and his team at JPL.

Horowitz studied the Cameron samples using both combustion analysis and a gas chromatograph–mass spectrometer (GCMS). The combustion analysis showed very tiny amounts of combustible material, but the GCMS indicated that none, or practically none, of this represented true organic compounds. Horowitz therefore concluded that the combustible material was elemental carbon, bits of coal dust transported to the Antarctic by wind. Horowitz also cultured some of the samples in conventional culture media, in a few cases obtaining colonies of well-known temperate-zone bacteria. But there were no such colonies evident in the soil samples themselves. His conclusion was that there might be a few bacteria in the soil, but these, like the coal dust, represented material that had been transported to Antarctica by the wind and had existed in the Antarctic soil in dormant form. As far as actually supporting life, Horowitz said, the

Cameron samples of the Antarctic dry valleys were sterile. These results were published by Horowitz, Cameron, and Horowitz's JPL coworker Jerry Hubbard in *Science* magazine in 1969[3] and 1972.[4]

In their 1972 paper, Horowitz, Cameron, and Hubbard wrote:

> The Antarctic desert is far more hospitable to terrestrial life than is Mars, particularly in regard to the abundance of water. In other respects too—such as the ultraviolet flux at the surface—Mars is decidedly more hostile than is the Antarctic.
>
> Second, Martian life, if any, must have evolved special means for obtaining and retaining water. . . . This has been known for some time. What is new in these findings is that even under severe selective pressure microbial life in the Antarctic has been unable to discover a comparable mechanism. To some this may suggest that life on Mars is an impossibility. In view of the very different histories of Mars and the dry valleys . . . we believe that such a conclusion is not justified.
>
> Finally, the Antarctic has provided us with a natural environment as much like Mars as we are likely to find on Earth. In this environment, the capacity of life as we know it to adapt and survive is pushed to the limit. The concentration of living things around sources of water in the dry valleys and their rapid drying out in the most arid locales may be useful as a model of the distribution of the life we may, if we are lucky, find on Mars.

The above excerpt indicates not self-contradiction but more likely a reasonable diversity of opinion within the Horowitz, Cameron, Hubbard team. The first paragraph and most of the second say that Antarctica is like Mars, but nicer, and life has been unable to evolve mechanisms allowing it to survive there. The obvious conclusion from this is that life can't make it on Mars. This would be appear to be Horowitz speaking, as he is widely known as a pessimist regarding the possibility of life on Mars. The team as a whole appears to have lacked consensus on this issue, however, as shown in the final sentence of the second paragraph, where the authors explicitly deny that they are willing to draw this inference as a conclusion. Then, in the third paragraph, the authors state that studying how life may adapt to Antarctica (which their paper says it has not managed to do) may be useful in revealing to us the modus operandi adopted by life on Mars.

The last paragraph is also a concise statement of the thesis behind Mars analog research. We can find out about how life would live on Mars by looking at how it does in terrestrial environments that mimic Mars. Given that Cameron was the one who put the most into the project, my guess is that he was the author responsible for it. A person does not usually engage in three years of heroic fieldwork in the hope of generating a negative conclusion.

But despite the caveat at the end of the second paragraph, a negative conclusion it certainly was. As a strong believer in the prevalence of life in the universe, Vishniac was unwilling to accept it.

In 1972, Vishniac, together with his graduate student Stan Mainzer, traveled down to Antarctica for the first time. They brought with them three methods of life detection: the Wolf Trap, conventional glass microscope slides, and a field version of an instrument called Gulliver. Gulliver was the baby of another Viking scientist, Gil Levin of the Baltimore-based Biospherics Inc. research firm. A fellow optimist on the life-on-Mars question, Levin was a bit luckier than Vishniac—his instrument, evolved somewhat and renamed the Labeled Release Experiment, actually eventually flew to Mars when the Viking spacecraft finally lifted off in 1975. Gulliver worked by exposing soil to a small amount of radioactive nutrient, and then counting the number of radioactive carbon dioxide molecules released from the soil as a result of microbial metabolism of the nutrient. According to Levin, the technique could count as few as ten living cells in a soil sample, which would make it about 10 million times more sensitive in detecting life than the Viking GCMS.

Vishniac and Mainzer journeyed to the same dry valleys that had been visited by Cameron, and tried out all their techniques. Both the Wolf Trap and Gulliver gave positive data strongly supporting the hypothesis of extant life in the soil. But the most fascinating results were obtained using the old-fashioned glass slides. Vishniac's technique for using these was a marvel of retrograde simplicity. He would simply stick dozens of the sterilized slides into the ground all over the place, and then come back a week later to collect them for examination under his microscope. In not only a few cases, but in fact the majority of them, Vishniac found microcolonies of bacteria—sometimes thousands of microcolonies—on his slides when he retrieved them from the soil. Since there had been no microcolonies on the slides when he put them in the ground, this proved that the bacteria in the soil were not dormant chaff thrown over from Australia but actual life living and multiplying in the Antarctic soil itself. In addition to the bacteria, Vishniac and Mainzer also found fungi, algae, and diatoms.

Returning with their samples to New Zealand, Vishniac and Mainzer showed them to John Waid of the University of Canterbury in Christchurch. Waid, an expert in electron microscopy, was able to image the sampled microbes at much higher magnification than Vishniac and Mainzer had been able to do in the field with their optical microscope. Waid's observations showed that some of the Antarctic bacterial colonies found by Vishniac exhibited unique adaptations. For example, some of them were in colonies encapsulated by a kind of shell "reminiscent of a honeycomb or a sponge."[5] The shell appeared to have the characteristics required to protect the encased bacteria from desiccation and ultraviolet light—exactly what would be needed by cells native to Antarctica but which would not be expected in temperate-zone microbes blown in from overseas.

Vishniac's results were a slap in the face to Horowitz, who by then was a powerhouse in the NASA Bioscience Section. He did not accept Vishniac's findings, saying that the slides must have been contaminated. Possibly as a result of this skepticism *Science* magazine refused to publish Vishniac's paper.[5] His only recourse was to go back to Antarctica and get more data.

But aside from Horowitz's skepticism, there was another problem with Vishniac's 1972 results. While he had, at least to his own satisfaction, shown that life could, and did, live in the Antarctic dry valleys, he had not found anything that he could reasonably claim was sufficiently well adapted to extreme conditions to actually be able to survive on Mars. During the spring of 1973, however, Vishniac was contacted by Imre Friedmann, of Florida

State University. Friedmann, an Israeli, had previously done research in the Negev Desert, where he had discovered that cyanobacteria and algae live under the surface of dry hot desert rocks. Friedmann, who lacked the funds to go to Antarctica himself, gave Vishniac a specimen from the Negev and asked him to look for comparable types of rock samples for him. Perhaps similar or even more rugged organisms could be found in the more extreme environment of Antarctica.

So preparations went forward for a return trip to Antarctica. Vishniac decided to take with him Zeddie Bowen, a professor of geology at the University of Rochester, where Vishniac also taught. Bowen's role would be to determine rock types, which would be especially important given Friedmann's research interests. In the summer of 1973, however, disaster struck again. Vishniac failed his physical examination. He had polio as a boy, and his right arm was shorter than his left. In the opinion of the Navy medical examiners, the problem was too serious to allow him to return to Antarctica.

But Vishniac was not about to give up. Furthermore, being Vishniac, he was not without recourse. His father was a world-renowned photographer. His wife, the brilliant microbiologist Helen Simpson Vishniac, was the daughter of the famous biologist and author George Gaylord Simpson. The family had connections. They knew senators. One of the senators pulled some strings, and the Navy was overruled.[5]

Thus it was that in September 1973, Vishniac and Bowen reached Antarctica. For the first two months they based themselves at McMurdo Station, collecting rocks and studying the water-absorption characteristics of the soil particles. Then in November, they moved their base camp to Mount Baldr, in the Asgard Mountain range. Vishniac wanted to get samples from the windblown mountain slopes, because this was the most hostile area for life, and therefore of greatest relevance to Mars.

That brings us to December 10, 1973. Vishniac had placed a number of glass slides between Mount Baldr and Mount Thor; he wanted to retrieve them for examination. Bowen, however, had work to do at base camp. That was not a problem, Vishniac replied, the trail was well marked. He would be back in twelve hours.

Twelve hours went by, but Vishniac did not return. Bowen waited four more, as agreed between the two, then set off to search for his companion. He found Vishniac's body at the base of a cliff. Apparently the biologist had found something of interest, and strayed off the premarked trail. Then he slipped on a steep slope and fell, tumbled, and bounced more than 150 meters to his death.

Vishniac's personal effects were flown out of Antarctica and returned to his widow Helen, in Rochester, New York. A microbiologist herself, she studied the soil samples and discovered a new species of yeast. But there were some rocks in his trunk labeled "For Imre." She sent these on to Imre Friedmann.

Friedmann cracked the rocks open, and under the translucent surface of some of them he found living communities of microorganisms. The translucent rock surface acted like greenhouse walls, shielding out deadly ultraviolet light and holding in heat and the tiny drop of subsurface liquid water that the heat produced. Beneath each tiny greenhouse wall was a miniature aquarium filled with cyanobacteria, algae, and fungi, all living happily in the coldest and near-driest place on Earth. Friedmann called the protected communities cryptoendoliths, meaning lifeforms hidden inside of rocks. Friedmann realized instantly that such communities might be entirely capable of living on Mars today. The field of exobiology finally had something to study.

Carl Sagan wrote Vishniac's obituary in the astronomer's journal *Icarus*. Vishniac, he said, was the second man to die for the idea of extraterrestrial life. The first was Giordano Bruno, murdered by the Inquisition in Rome in 1600.

THE VIKINGS LAND

After a delay of two years, the two Viking spacecraft were finally launched in August and September 1975. The launches were timed to allow Viking 1 to land on July 4, 1976, the American bicentennial. But while Viking 1 made it to Mars orbit on schedule, a proper landing site could not be found in time, and the plan for a bicentennial-bash Mars landing had to be abandoned. Things worked out for the best, however, as the actual landing took place on July 20, just seven years to the day after the Apollo 11 Moon landing. So instead of being used to celebrate an old holiday, the Viking landing was used to help create a new one. July 20 is Space Day.

The Wolf Trap was gone, but there remained three life-detection experiments on Viking. All worked on variations of the same general principle: a culture medium of some type would be exposed to Martian soil, and then under various conditions it was expected that organisms, if present, would cause gases of one type or another to be emitted. By measuring the amount of emitted gas, the scientists could assess the presence or absence of life.

Between the eighth and eleventh day after the Viking 1 landing, all three life-detection experiments were exposed to Mars dirt, and incredibly, all three registered large gas releases. This was just what the experiment designers had hoped for: impressive, unambiguous gas releases. According to the plan of the experiment, this response signaled the presence of life. But then something unexpected happened: the gas emissions stopped. Now the unambiguous gas emissions suddenly took on an ambiguous meaning. Were the gas releases caused by life, or by a chemical reaction with elements in the soil? Finite gas releases suggested the latter explanation—chemicals react until they are gone and then the reaction stops. But a biological explanation is not impossible either. Perhaps upon being exposed to the bounty of the Viking experiment culture medium, the Marsbugs multiplied out of control, producing a strong gas-emission signal for a while, only to stop when they poisoned themselves in their own wastes.

To attempt to resolve this question, Viking had one remaining experiment, a Gas Chromatograph Mass Spectrometer (GCMS) capable of detecting the presence of small amounts of organic material in the soil. It found none.

For the life-pessimists on the team, such as Norm Horowitz, that was enough. The Viking gas releases were chemical reactions between the culture media and peroxides in the soil. End of story. For the life-optimists, such as Gil Levin, this was nonsense. The experiments were designed to indicate life by gas release, and that is exactly what they did. The GCMS negative result was irrelevant. It had, after all, failed to find life in soil where Vishniac found plenty.

And that is where the matter stands today. The jury is split, and there is insufficient evidence to decide the matter. Would we be substantially better off if the Wolf Trap had flown? I'm not sure. In the context of Viking's other results, either a positive or negative response by the Wolf Trap could easily have been explained away. If the experiment's liquid had become cloudy, Horowitz would have dismissed it as a chemical response. If it had remained clear, Levin would have argued that the Marsbugs drowned.

If you want to find life on Mars you don't need $30 million automated instruments. You need a $500 microscope, a large package of ten-cent glass slides, and someone there who really knows how to use them.

EXPANDING THE SEARCH

But we don't have anyone on Mars yet. So if we want to use the unique strengths of human explorers, we need to use them to explore Mars on Earth. With the death of Vishniac, the torch for leading this kind of work was passed to Imre Friedmann. He continued and expanded his work on endolithic organisms, discovering many new types in various cold and hot deserts all around the world.[6] As the years went on, he transformed the study of Mars analog environment organisms from a personal specialty into a field of research covering a broad range of issues and environments and mobilizing the talents of numerous devoted scientists. One of the people he recruited to the field was Chris McKay, who joined him on an expedition to Antarctica in 1980. The two became collaborators, working together to uncover the endolithic secrets of many new Mars analog environments, including Siberia, the Gobi and Atacama Deserts, and the Canadian far north.

In Siberia, McKay worked with the Russian team led by David

Gilichinsky, which had found endoliths frozen dormant millions of years ago, but which were still viable. Such ability to go dormant for very long periods of time could be key to the life cycle of Martian endoliths, whose ability to transfer and spread from one rock to another might be limited by very intermittent periods in which the outside environment was sufficiently favorable for life. Such intermittent favorable periods could be created either by occasional outflows of water from Mars's interior, meteoric impact, volcanic activity, or by the motion of Mars's rotational axis, which wobbles over wide variations on a time scale of about 100,000 years, causing major climatic changes in the process.

In Antarctica, McKay dove beneath the surface of permanently ice-covered lakes to find active ecologies maintained by warming from incoming snow meltwater, despite an ambient annual mean temperature of –20°C. Such lakes could provide an analog for potentially life-bearing environments on early and adolescent Mars. In Chile's Atacama Desert, Earth's most arid locale, McKay, moving in from the coastline, found a termination zone beyond which not even endolithic algae could survive. Such zones, possessing less than a millimeter of precipitation per year, could define moisture boundary that needs to be at least locally crossed if we hope to find life on the Martian surface.

Among his later discoveries, Friedmann found a kind of limestone-boring endolith in the Negev, which eats carbonate rock, eventually causing its carbon content to be released as carbon dioxide. Possibly under the influence of McKay, whose central interests include Mars terraforming, Friedmann proposed using such bugs to thicken Mars's carbon dioxide atmosphere and thereby bring the planet to life via greenhouse warming.

Other researchers who joined the analog Mars exploration effort included Jack Farmer, who focused his attention on hydrothermal springs, and Penelope Boston, who devoted herself to researching the unique biological systems that exist in deep caves. Both of these types of environments may exist as niche oases supporting life on Mars.

MARS COMES TO EARTH

But we are not entirely limited to studying analogs of Mars on Earth. A few pieces of the Red Planet have taken the trouble to travel here. In 1996, one of them, a 2-kilogram rock known as ALH84001, caused quite a stir.

The history of ALH84001 is as follows. The rock was formed a kilometer or two underground on Mars about 4.5 million years ago, not long after the formation of the planet itself. About 3.6 billion years ago it was fractured, probably by a meteor impact event not too far away from its location. Then, about 16 million years ago, another impact event ejected it from Mars, causing it to fly around in space, until as chance would have it, it encountered the Earth 13,000 years ago and landed in Antarctica. All of these facts are known through various kinds of chemical analysis and isotopic dating techniques. For example, the rock's Martian origin is proved by its oxygen isotope ratios and the discovery of pockets of encased gas whose composition matches that of the Mars atmosphere as measured by Viking. Its formation date is supported by both samarium-neodymium and rubidium-strontium ratios, which allow the determination of age because these are mother-daughter radioactive decay pairs. Similar mother-daughter isotope ratios indicate the date of the fracturing impact, although with less certainty. The time it spent on Earth after landing can be determined by conventional carbon 14 dating, and the time spent during spaceflight is known from the amount of isotope changes induced by cosmic rays. Adding these two latter amounts of time together gives us the date of the planetary ejection event. Thus the basic chronology of ALH84001's career is not in dispute.

National Science Foundation Antarctic Meteorite program geologist Roberta Score found the rock in Antarctica's Allan Hills in early 1984 (thus the name, ALH84001). Despite being identified by Score immediately as anomalous, the rock was placed in cold storage at NASA Johnson Space Center and more or less ignored. There it remained until 1993, when a sample of it was mistakenly delivered to meteorite researcher David

Mittlefehldt, who had ordered a different rock for examination as part of his program of studying diogenites. Mittlefehldt saw zoned carbonates in the rock and realized ALH84001 was no ordinary meteorite; it was a member of the one-in-a-thousand class that comes to Earth from Mars.

Thus rescued from obscurity by a process somewhat analogous to the workings of the judicial system of the old Austro-Hungarian Empire ("despotism tempered by inefficiency"), ALH84001 was made the subject of special study by a team of researchers.

By August 1996, the team, consisting of Johnson Space Center scientists David McKay, Everett Gibson, Kathie Thomas-Keprta, and Chris Romanek, and Stanford University Chemist Richard Zare, had written up their study in a *Science* magazine[7,8] paper showing remarkable results. According to the paper, ALH84001 showed evidence indicating the presence of bacteria on Mars 3.5 billion years ago. A discovery of this magnitude had obvious political implications, and so the team briefed NASA Administrator Dan Goldin. Goldin thereupon briefed Vice President Al Gore, who gave copies of the preprinted paper to a White House political strategist, whose mistress leaked them to the press. Thus a secret that had been kept by a fairly substantial group within the scientific community for over a year (I, who had no involvement whatsoever in this work, heard a little about it as early as the summer of 1995) was leaked by the White House inside of a week. To preempt distorted press coverage, the team had no choice but call a press conference explaining their discovery prior to formal publication.

The press conference was held at NASA headquarters on August 6, 1996. Introduced by NASA Administrator Goldin, the presenting panel included McKay, Gibson, Thomas-Keprta, and Zare, as well as noted paleobiologist J. William Schopf, the Director of UCLA's Center for the Study of Evolution and the Origin of Life, who was there to provide skeptical counterpoint.

As presented, the team's lines of evidence arguing for past biological activity included the presence in the rock of carbonate globules, organic polycyclic aromatic hydrocarbons (PAHs), photographs of small structures resembling fossil bacteria, and crystals of minerals including pyrrhotite, greigite, and magnetite that are frequently created biogenically. We briefly address each of these types of evidence in turn.

Carbonates: Carbonates are formed by the reaction between carbon-dioxide-bearing water and rocks. There is no doubt that carbonates exist in ALH84001. This is evidence not for life per se but for the existence of an aqueous environment that could have supported life. The team argued that their evidence showed that the carbonates formed at temperatures below 80°C (176°F), which would be acceptable for life. Skeptics have proposed alternative mechanisms whereby the carbonates could have formed at temperatures as high as 450°C (842°F), perhaps during one of the impact shock events. The greigite and PAHs in the rock would have been destroyed at such temperatures, however, so this high-temperature explanation is inconsistent with the data. Furthermore, discounting the carbonates in ALH84001 seems beside the point. We know from orbital imaging of water-erosion features that there was liquid water on Mars 3.5 billion years ago, and it was almost certainly well loaded with carbon dioxide. It would thus be surprising if a Mars rock from that period did not contain carbonates. In any case, the existence of the aqueous environment in ancient Mars that the carbonate is being used to demonstrate is beyond dispute.

PAHs: PAHs are organic molecules, but that does not mean they were created by life. They have been found in ordinary meteorites, and few would argue for a biological origin in those cases. In addition, some have dismissed the PAHs in ALH84001, arguing that they could be the result of terrestrial contamination. The PAHs in ALH84001 do not represent the full spectrum of PAHs found in terrestrial pollution, and they are present at about a thousand times the concentration ordinarily found in Arctic or Antarctic samples that have been contaminated by Earth's atmosphere. Moreover, the concentration of PAHs in ALH84001 increases as one moves from the exterior to the interior, which is the exact opposite of what one would observe if the source of the PAHs were terrestrial contamination. So the PAHs are from Mars, but they don't prove life. They do show, however, that there was organic chemistry going on underground on Mars at that time, which is very interesting.

Fossil-like structures: The team displayed electron microscope photographs of things that certainly looked like fossils. One even had an uncanny resemblance to a segmented worm. There are two problems with this line of evidence, however. In the first place, as Schopf pointed out at the press conference, inorganic processes can frequently produce little rock structures, called "foolers," that look like fossils but are not. The second problem is that the fossil-looking things in ALH84001 are about an order of magnitude smaller than any known bacteria. There is thus a basis for arguing that they cannot be fossils, because there is no way that all the biochemical complexity that goes into making a bacterium can be packaged in a volume that small. I don't think that this argument is valid; it's like saying that an unprecedentedly large fish cannot be a fish because no previous fish was ever that big. Moreover, for reasons that I will explain below, I think that searching for lifeforms, or fossils of lifeforms, smaller and simpler than known bacteria is exactly the most important form of exobiological research to be done on Mars. So ruling such things out if we see them is a bad idea. That said, however, there is no proof that the fossil-like formations observed in ALH84001 are in fact fossils.

Possible Biogenic Minerals: As noted above, the team showed that ALH84001 contained a number of sets of tiny crystals of various minerals, including magnetite and pyrrhotite, that generally are found on Earth as products of bacterial activity. Unfortunately, they can also be produced by inorganic processes. The team showed that some of the carbonate in the area where the minerals were found appeared to have been eroded by acids, which would indicate the prevalence of pH conditions incompatible with the basic chemistry required to precipitate magnetite and pyrrhotite by inorganic means. Schopf, however, was unconvinced. The two events could have happened at different times. If you want to prove that such minerals are of biologic origin, he said, you need to show that they have been laid down in the kind of linear formations, or chains, that are unique to life. The team had not done that.

Thus, while they had identified four intriguing types of phenomena friendly to biology, none of these individually proved the existence of past life in ALH84001. The team argued however, that taken in combination, the simplest explanation for the ensemble was biological activity. But preponderance of evidence is a subjective criterion, and speaking for many scientists of similar temperament, Schopf was unconvinced. Quoting a famous dictum attributed to Carl Sagan, he said, "Extraordinary claims require extraordinary evidence."

So there the battle lines were drawn, with the team and others gathered to their side attempting to expand or defend the evidence, and large numbers of skeptics attempting to debunk it. Much of this argument generated more heat than light. For example, at the Mars Society convention in Toronto in August 2000, there was a one-on-one debate between Simon Clemett, a coworker of chemist Richard Zare and a member of the team, and Ralph Harvey, of Case Western Reserve University, an outspoken skeptic. Harvey was the far more capable debater, and appeared to get the better of Clemett, but many of his points were invalid. For example, Harvey derided the fact that the team had only chosen to publish the photographs of the rock that contained the apparent microfossils, but had not published the much larger number that failed to show any such objects. This was nonsense. If you want to show that deer live in the woods you only need to show the photos in which deer appear. Discarding the null pictures is entirely legitimate.

One of the positive aspects of the dispute, however, was that it resulted in the most intense study of any single meteorite, Martian or otherwise. As part of these investigations, Joseph Kirschvink of Cal Tech made an important discovery; he found chemical evidence showing that in the course of its journey from Mars ejection to Earth landing, large parts of the rock had never exceeded 40°C (104°F).[9] This confirmed an earlier theoretical calculation by Jay Melosh of the University of Arizona, who had predicted, based on the mathematics of shock interactions, that rocks could be ejected from planets without being excessively heated.[10] Kirschvink's experimental confirmation of Melosh's math is very important, because it means that material can be transferred between planets *without being sterilized.* Thus if there had been bacteria in ALH84001 at the time of its ejection, they would have survived the trip to Earth.

Of course, ALH84001 arrived on Earth only 13,000 years ago, and any Marsbugs riding on it would have been greeted by hordes of native Earthling bacteria, fully adapted to their homeworld's environment and eager to eat the awkward newcomers. But what of the more distant past? Mars rocks have been landing on Earth (and Earth rocks on Mars) since the birth of the solar system. Mars, being smaller, cooled from its initial molten state before the Earth did. Life would have had its chance to get started on Mars before it did on Earth. So what if Martian bacteria were delivered to Earth before there were any natives to challenge them? *What if life on Earth came from Mars?* Kirschvink's discovery threw that possibility wide open.

Then, in the fall of 2000, a bombshell was dropped into the debate. The ordnance in question came from none other than Imre Friedmann, getting on in years, but as events would show, still kicking. Hard.

Schopf had dismissed the original ALH84001 team's magnetite claims because they had found no linear formations, or chains. Magnetite crystals created by magnetotactic bacteria are laid down in chains. In contrast, magnetite crystals created by nonbiological processes have no such geometrical organization.

Friedmann found the magnetite chains. They were there all right, and in a paper published in February 2001 he and his collaborators showed them for all the world to see.[11] Not only that, the Friedmann team (consisting of Friedmann, Jacek Wierzchos of the University of Lleida, Spain, Carmen Ascaso of the Centro de Ciencias Medioambientales, Madrid, and Michael Winklhofer of the Geophysics Institute of the University of Munich) showed they met a set of criteria that very forcefully indicated a biological origin. These criteria "which could not be present in abiotically formed chains of magnetite crystals (no such chains have ever been observed in nature)," were (1) uniform crystal size and shape within chains, (2) gaps between crystals, (3) orientation of elongated crystals along the chain axis, (4) halo traces of membrane around chains, and (5) flexibility (bending) of chains. Friedmann et al. didn't mince words: "We conclude that the chains of electron-opaque particles in ALH84001 are magnetofossils, as no other consistent explanation would account for these findings."

Despite denials from the other side, Friedmann et al. really did nail the lid on the Alan Hills meteorite controversy. If extraordinary claims require extraordinary evidence, they had certainly supplied it. In light of Kirschvink's proof of the possibility of interplanetary bacterial transfer, however, by 2001 showing the existence of life in ALH84001 was really no extraordinary claim at all. After all, we know there was life on Earth 3.6 billion years ago, at a time when there was liquid water on Mars. Furthermore, with the higher rate of asteroidal impacts typical of the early solar system, there was certainly plenty of natural transport available. So there *must* have been some bacteria on Mars at that time, if from no other source than the *Earth.* The real question is what the source of the bacteria was. We shall return to the significance of this question shortly.

But by validating the magnetofossils in ALH84001, the Friedmann team did much more than show the presence of life, they showed the presence of a particular form of life, magnetotactic microbes. Now, on Earth magnetotactic bacteria use their little compasses to allow them to navigate up and down to reach places where the oxygen concentration in their fluid medium suits them best. Therefore, one does not find magnetotactic bacteria on Earth until the oxygen level in the atmosphere had built up to significant concentrations, roughly 2.3 billion years ago. Readers who know some geological history may wonder about this; after all, it is well known that photosynthetic cyanobacteria appeared on Earth close to 3.5 billion years ago. Why did it take so long for our planet to start to become oxygenated? The reason is that the limited amount of photosynthesis the primitive cyanobacteria could perform could not overpower the capacity of the Earth's plate tectonics to recycle fixed carbon back into the atmosphere as carbon dioxide.

Friedmann's findings thus imply the presence of free oxygen on Mars in significant quantities more than a billion years before it was available on Earth. This is not too surprising. Because it is a smaller planet, Mars's tectonic activity is much weaker than the Earth's—in fact, it is essentially nonexistent today. Therefore, the Red Planet would not recycle biologically fixed carbon as effectively as the Earth, and this might give primitive cyanobacteria a chance to oxygenate the place much faster.

But here is where things get wild. On Earth there is significant evidence that the rate of evolution is correlated with the concentration of oxygen in the atmosphere. There is a clear statistical correlation between these factors, but there is a logical causal relationship as well. The presence of oxygen allows for more energetic chemical reactions, and thus more energetic and complex organisms. If we consider the development of animals, for example, these complex oxygen-breathing creatures are each composed of multicelluar organizations of oxygen-respiring nucleated cells. These nucleated cells, or eukaryotes, are themselves complex organizations of subsystems, such as mitochondria (cellular energy generators), which in the distant past were once free-living bacteria. According to the now generally accepted theory known as symbiogenesis, developed originally by biologist Lynn Margulis of Boston University, it is believed that the complex nucleated cells that compose all higher animals and plants themselves originated as colonies of bacteria whose different members evolved to specialize in various activities. Thus bacteria have an analogous relationship to animal (or plant) cells, as single-celled animals do to multicellular animals.[12]

Through examination of the fossil and geologic record, it has been determined that the appearance of cells using mitochondria correlates with the rise of atmospheric oxygen concentration to between 1 percent and 2 percent of present atmospheric levels (PAL). Chloroplasts (specialized cellular photosynthesis units) appeared roughly 2 billion years ago, when oxygen levels rose to 5 percent PAL. Around 600 million years ago, with oxygen levels rising to about 20 percent PAL, multicellular animals burst upon the scene with a suddenness that has caused their advent to be called "the Cambrian Explosion."

The significance of the relationship between atmospheric oxygen levels and Martian evolution was first identified by astrobiologist Chris McKay in a series of daring papers published in 1996.[13, 14] In these papers, McKay argued that we should not take the pace of evolution on Earth as a necessarily representative model. The Earth took 3.2 billion years from the end of the heavy asteroidal bombardment (which presumably precluded any life before its conclusion) period 3.8 billion years ago to generate multicellular life, but since the rate of evolution is conditioned by the presence of oxygen, on Mars things conceivably could have occurred much faster. Mars's warm-wet juvenile period only lasted about 1 billion years before the CO_2 atmosphere thinned out and the planet lost its beneficial greenhouse effect. On Earth, evolution acting over such a span was only able to yield bacteria. On Mars, with greater amounts of free oxygen present, it might have produced much more. It might have produced nucleated cells. It might even have produced complex multicellular animals and plants.

In 1996, when McKay proposed these ideas, most people, including me, regarded them as fanciful speculations. But Friedmann's demonstration of Martian magnetotactic bacteria changed that. He had shown that on Mars, just 200 million years after the end of the heavy bombardment, there were critters of a type that took 1600 million years to appear on Earth. Suddenly McKay's ideas didn't seem so weird after all.

MARS AND THE ORIGIN OF LIFE ON EARTH

The origin of life on Earth is a mystery. Despite centuries of investigation by innumerable researchers, no evidence has ever been produced for the presence on Earth, either now or at any time in the past, of any free-living microorganisms simpler than bacteria. This is a striking fact. While bacteria are frequently thought of as simple life forms, they are actually extremely complex molecular machines, using highly evolved mechanisms to enable survival, metabolism, growth, reproduction, mobility, and innumerable other functions. It is thus inconceivable that bacteria could actually represent the earliest life to emerge from chemistry. A period of prior evolution had to occur, starting with much simpler forms, developing by degrees into the elaborate organisms we call bacteria. Yet we have good fossils of cyanobacteria, apparently similar to current forms, existing on Earth 3.5 billion years ago. This is only 300 million years after the close of the heavy bombardment period that would have made life impossible here prior to 3.8 billion years ago. This is an extraordinarily short time to expect native bacteria to evolve from chemistry, especially if we consider that the fossil record shows that for the next 2 billion years, the pace of evolution on this planet was extremely slow.

From the mathematical point of view, it is apparent that the general pace of evolutionary change in the biosphere is fastest at the present, and the farther back we go in the past, as illuminated by the known fossil record, the slower evolution occurs. Thus it took 2 billion years for bacteria to evolve sufficiently to produce nucleated single-celled organisms (eucaria), but only another 900 million years to produce the first true multicellular plants and animals. In another 400 million years we see complex vascular plants, fish, amphibians, reptiles, and protomammals, and in the next 200 million years, we see trees with seeds, grasses, flowering plants, dinosaurs, birds, mammals, and man. As we have seen, this rate is correlated with the concentration of oxygen in the atmosphere. But the general pattern observed also is that the more evolved life becomes, the greater its capability develops for ever more rapid evolution. It thus appears to be rather a stretch to argue that the simplest life forms that preceded bacteria should have been able to span the huge evolutionary chasm separating organic chemistry from complex bacteria in a geological blink of an eye, but then have evolution slam on the brakes for the next 2 billion years. If anything, it is to be expected that such simple prebacteria operating in an oxygen-deprived environment should have accomplished their evolutionary ascent in the most tedious fashion.

Furthermore, as mentioned above, there are no surviving examples of this class of organisms on Earth. This seems extraordinary, and is not well explained by the supposition that such creatures were driven into extinction by the more highly evolved bacteria. After all, despite the development of the more complex eucaria, bacteria are still very much around, and the single-celled eucaria survive quite nicely despite the evolution of still more complex animals and plants. Complexity always comes at an evolutionary cost, leaving plenty of room for the simpler folk who preceded the more complex forms.

Thus, while bacteria could not have been the first life, both the fossil record and current biological surveys strongly support the assertion that bacteria were in fact the first life *on Earth*. The only way to resolve this situation is to hypothesize that bacteria did not evolve on Earth, but arrived here fully evolved from space. This hypothesis, known generally as panspermia, is further supported by the observation that many varieties of bacteria have adaptations that allow them to survive in dormancy for long periods in hard vacuum, ultracold, and radiation environments that can only be found in outer space. In general, in biology, all adaptations have costs, and organisms do not support adaptations that have no utility. If we found a species of land animal with vestigial aquatic adaptations, we would assume that its ancestors came from the sea. Similarly, it can be argued that the presence of astronautical adaptations among bacteria strongly supports the conjecture that their ancestors came from space.

The panspermia hypothesis is unpopular among origin-of-life researchers, because it completely ducks the central question of interest in their field, the origin of living things from nonliving chemistry. Actually, panspermia is not irrelevant to these concerns, because it opens up the possibility that life may have originated in more favorable environments than the early Earth—for example, on a planet offering the chemically reduced environment found so favorable for the development of amino acids in the 1950s experiments by Miller and Urey.[15] In these experiments, Harold Urey's graduate student Stanley Miller achieved scientific immortality by mixing methane, ammonia, and water vapor in a flask and zapping the mixture with electric sparks to produce a large number of the amino acids that are considered fundamental (and, prior to Miller, unique) to biology. These experiments have been criticized as irrelevant to the origin of life, because the early Earth offered an oxidizing environment in which the Miller-Urey reactions would not readily occur. If panspermia is possible, however, then these criticisms are moot. Whatever its relevance to questions of the origin of life, it is clear that the panspermia hypothesis bears heavily on the issues of the prevalence and distribution of life in the universe.

TRANSPORT OF LIFE BETWEEN EARTH AND MARS

As discussed above, it is now well established that throughout its history, the Earth has been the target of numerous impacts by asteroids and comets, which have had the capability to eject large amounts of unshocked, and therefore unsterilized, material into interplanetary and interstellar space. Collaborators of University of Arizona Professor Jay Melosh, such as Swedish biologist Curt Mileikowski, have published calculations showing that significant fractions of this material can find their way to nearby planets such as Mars within time scales that are very short compared with the demonstrated viable life spans of dormant bacteria.[8] Thus, throughout geological history, innumerable bacteria have without question been transported from Earth to Mars. Furthermore, if there is or ever was bacterial life on Mars, natural transport of these organisms has occurred from Mars to Earth as well. Indeed, 500 kilograms/year of unsterilized Martian rocks are estimated to fall on Earth every year. This observation indicates that the current very expensive "planetary protection" programs instituted by various space agencies to prevent the transportation of microorganisms between planets on artificial spacecraft are without rational foundation; the microbes already have plenty of their own spacecraft and have been making the trip regularly and in large numbers for the past 3.5 billion years.

The ease of natural transport of bacteria between Earth and Mars makes it unlikely that any past or present Martian life could have a separate origin from terrestrial life. For two separate origins to occur, they would have to be nearly simultaneous, as otherwise the planetary life that originated first would preemptively spread. Rather, the most realistic possibilities are that both Earth and Mars were seeded simultaneously from an outside, presumably interstellar source, or that life developed indigenously from chemistry on either Earth or Mars, and then rapidly seeded the other as soon as forms (such as bacteria) that could survive spaceflight evolved on the planet of origin. We have seen from the previous discussion that the lack of prebacterial organisms on Earth undermines the supposition that life developed indigenously on Earth. Therefore, the most probable alternatives are either (1) life originated on Mars and then seeded the Earth, or (2) the Earth and Mars were seeded simultaneously from an interstellar source.

It will be observed that I do not include as a possibility that only Earth, and not Mars, could have been seeded from the outside. This is because it is now clear that Mars had liquid water on its surface for hundreds of millions of years of its early history. Thus, if an interstellar source of bacteria had been available, Mars certainly would have been seeded too.

The narrowing of possibilities to (1) and (2), above, recasts the issues surrounding the search for life on Mars in a decisive way. The key point is not whether bacteria ever existed on Mars; they almost certainly did. Rather the key issue is whether prebacterial organisms (prebacteria) either do or did exist on Mars. If we can find evidence for such prebacteria on the Red Planet, then we should conclude that (1) is true. If we don't, then we must conclude (2).

The implications of either of these results would be spectacular. For example, if prebacteria can be found on Mars, then we would finally gain an understanding of the fundamental steps involved in the transition from chemistry to life. Furthermore, since much of Mars's surface is fairly well preserved going back 3.8 billion years, we would have the opportunity to read the history of life's development from nonliving chemistry directly from the fossil record. In essence, we would be gaining the opportunity to read the book of life itself.

On the other hand, if the search for life on Mars only reveals evidence for the same sort of well-developed spaceflight-capable bacteria we see as the earliest inhabitants of the Earth, then that would say that both planets were seeded from interstellar sources. That would prove the validity of interstellar panspermia, and therefore imply that microbial life should be expected on nearly all of the many billions of microbe-suitable planets that are now believed to exist throughout our galaxy. As microbes are the source for evolution to higher forms, this would greatly increase the odds for the widespread existence of complex and intelligent life as well.

The search for both living and fossilized prebacteria on Mars thus emerges as one central to an understanding of the place of both life and humanity in the universe. It seems pretty clear that this can only be undertaken competently by human explorers operating on the planet's surface. The reason for this is that prebacterial fossils, in the absence of distracting and more obvious bacterial fossils, are likely to be very old, and thus rare. Furthermore, if Chris McKay is right—and it now appears possible he might be—then we need to be looking for fossils of macroscopic animals and plants as well. Since Mars's wet period ended 3 billion years ago, any fossils of these creatures will have to be at least that old, and considerably more rare than dinosaur bones are on Earth. It would thus be wildly unreasonable to hope that any such things would be found in a kilogram of material returned to Earth as part of a robotic sample-return mission.

But the real scientific bonanza on Mars would come from the recovery of live organisms, either prebacteria, or if we are very lucky, bacteria or even eucaria of a separate origin. We need live organisms if we are to examine their structure, and learn what were the steps by which prebacteria made the transition from chemistry to life. We need live organisms if we are going to be able to determine for sure whether Martian bacteria represent a common or separate origin from those of the Earth. And if Mars life does represent a separate genesis, only living samples will tell us in what ways it was able to choose an alternative path from that followed by the life of the Earth.

To obtain such living samples, we will need to set up drilling rigs capable of penetrating a kilometer or more below the Martian surface to reach liquid ground water and the active biosphere it probably hosts. That is no task for small robotic probes.

If we are to find out the truth about the nature of life, human explorers will have to go to Mars.

3. ENTER THE MARS SOCIETY

> We may affirm absolutely that nothing great in the world has been accomplished without passion.
>
> —G. W. F. Hegel, *The Philosophy of History*

THE MARS UNDERGROUND

Boulder Colorado, 1978.

Apollo was over. The STG plan was long gone. The Shuttle was still scheduled to fly in a few years, offering supposed benefits to the nation's satellite launch capability. But as far as NASA's plans for human spaceflight were concerned, that was it. For the rest, the Viking mission had also run its course. Except for continued photography by the orbiters, and weather measurements on the ground, the mission was essentially finished, and despite protests by former Viking scientist Gil Levin and a few other mavericks, produced a fashionable consensus that the Red Planet was sterile. Since there were no plans for human planetary exploration, perceptions would stay that way. In fact, there were not even any plans for further robotic landers. Why bother? Mars was dead. Everyone knows it. Face the facts.

But there are some people who refuse to "face the facts"—who prefer to question them, or look behind them, or around them. People who are not interested in what everyone knows but devote themselves to thinking about what nobody knows. One of these people was future NASA scientist Christopher P. McKay.

In 1978, McKay was a graduate student in astrophysics at the University of Colorado at Boulder. He knew the Viking consensus said that Mars is dead. Well, maybe it is, reasoned McKay, but that is only the Mars of the present. There is also the Mars of the past, and the Mars of the future. The Viking orbiter's cameras had revealed large networks of dry riverbeds on the Martian surface. The Mars of the past had been warm and wet, a place friendly to life. Perhaps the Mars of the future could be so as well.

McKay wanted to examine that question. So he started a graduate seminar on terraforming Mars.

The word "terraforming" means planetary engineering with the specific intent to transform a sterile world into a living one. On Mars, for example, this could in principle be done by mass-producing artificial greenhouse gases to warm the planet to outgas its adsorbed CO_2 atmosphere and mobilize its frozen reserves of liquid water. The spread of oxygen-creating plant life across the surface would then become possible. The term "terraforming" was introduced by science-fiction writer Jack Williamson in the late 1940s, but the core idea behind it is much older. In fact, it goes back to the ancient Judeo-Christian notion that since man is the image of God, man should attempt to continue God's work. Terraforming is continuing the work of creation.

So whether aware of it or not, would-be terraformers engage in a design project to build a planetary cathedral celebrating the divine nature of human reason. Such discussions can be very inspiring.

McKay's terraforming seminar was no exception. It engaged and galvanized a group of some twenty-five graduate students and others with the intellectual challenge of bringing a dead world to life. At the core of the group were Carol Stoker, a fellow astrogeophysics graduate student; Penelope Boston, an undergraduate biology major and friend of McKay's; Tom Meyer, president of his own engineering firm and a close friend of Stoker's; and computer scientist Steve Welch, who later became Boston's significant other. Also drawn to the group was Carter Emmart, then only a high school student, but one possessed of formidable artistic talent. Emmart was able to take the group's ideas and render them into quick sketches or detailed paintings, thereby enabling the group to actually see and show others what they were talking about. Charles Barth, the director of the Laboratory for Atmospheric and Space Physics at the University of Colorado, acted as mentor and counselor to the group, helping them focus their discussions into a formal seminar, "The Habitability of Mars."

Over the course of the first semester, the seminar's participants, with some gentle nudges from Barth, recognized that terraforming Mars was a tall order, even for graduate students. They also realized that they were theory-rich and data-poor. While entertaining and intriguing, discussions of terraforming Mars would lead nowhere without more data. The group needed more information about Mars—its present atmosphere, its past atmosphere, volatiles, resources, a multitude of items—data that human missions could collect. So they began to focus on near-term human missions to Mars, and eventually wrote up their findings as "The Preliminary Report of the Mars Study Group." Barth shepherded the report to NASA headquarters, and word soon got out that a band of graduate students and others out in Boulder were enthusiastically—and intelligently—investigating human missions to Mars (and terraforming as well).

As the group shifted its emphasis from terraforming to human Mars exploration, they increasingly drew into their orbit, and themselves became influenced by, more established Mars researchers. One of these was Benton Clark. Ben Clark had been the principal investigator on the Viking X-ray Florescence experiment, which identified the inorganic chemical composition of the Martian soil. In 1978, observing the ambiguity of the Viking life-detection experiments, he had written a seminal essay entitled "The Viking Results—The Case for Man on Mars." A key section of his essay reads:

> The ultimate scientific study of Mars will be realized only with the coming of man—man who can conduct seismic and electromagnetic sounding surveys; who can launch balloons, drive rovers, establish geologic field relations, select rock samples and dissect them under the microscope; who can track clouds and witness other meteorological transients; who can drill for permafrost, examine core tubes, and insert heat-flow probes; and who, with his inimitable capacity for application of scientific insight and methodology, can pursue the quest for indigenous life-forms and perhaps discover the fossilized remains of an earlier biosphere.[16]

These ideas became a rallying cry for the group.

During the spring of 1980, McKay and Boston crossed paths with Leonard David at an American Astronautical Society meeting in Washington, D.C. David had spent the past few years arranging student forums on space exploration and had heard about the Boulder crew. The three hit it off fairly quickly, and what started as a chat about Mars exploration ended with David suggesting that some effort ought to be made to hold a conference on human Mars exploration. This was a novel idea, as twenty-something graduate students usually didn't organize and host planetary-exploration conferences. But adopting something of a "why not?" attitude, McKay, Meyer, Stoker, Boston, and Welch began some low-key planning. Via some graduate student guerrilla methods, they ran off a hundred or so copies of a conference announcement and bundled them off for distribution. Much to everyone's surprise, calls started coming in, both from those who wanted to attend and from researchers interested in delivering papers. Deriving the forum's name from Clark's celebrated article, in late April 1981, the Boulder group hosted the first "Case for Mars" conference.

It was a small conference—just a hundred or so people eventually attended, but to the organizers, these were legions. It was heartening, thrilling, indeed liberating to know that there were others who shared their passion. Leonard David had arrived from Washington with a bundle of red buttons. Imprinted on the buttons, below a Case for Mars logo depicting Leonardo da Vinci's classic drawing of Man enclosed within the circle of the ancient symbol for Mars, were the words "Mars Underground." Over the course of four days, numerous workshops, and a slew of presentations, the Mars Underground formulated plans for the human exploration of Mars: the whys and wherefores of the program; precursor missions to human missions; mission profiles; and rosters of surface activities for explorers.

The conferences continued, one every three years, each building on what preceded and reflecting the character of the times. The 1984 conference resulted in a complete end-to-end design for a Mars mission that underground members used as the basis for a two-hour presentation on Mars exploration that they delivered at NASA headquarters and other NASA centers. The 1984 conference was also notable in that it drew to the group people of greater political influence, such as former NASA Administrator Thomas Paine. In 1985, Paine managed to get himself appointed by President Reagan to head the blue-ribbon National Commission on Space, and then proceeded to guide it to a recommendation that the United States make the establishment of a human outpost on Mars the thirty-year goal of the space program. President Reagan responded to the report by setting up the Code Z organization and Pathfinder programs at NASA headquarters, to respectively plan mission strategies and develop the key technologies required for human expansion to the Moon and Mars. It was these organizations that formed the insider network that provided the policy input that led to the first president George Bush's call for a Space Exploration Initiative in July 1989.

The third Case for Mars conference accelerated the trend, with Carl Sagan giving the keynote to an audience of over a thousand people, including a substantial representation of the international press. I had first heard about the Mars Underground after Case for Mars II, and along with more than four hundred other out-of-town technical types, went to Case for Mars III to participate in some of the nearly two hundred presentations and sixteen workshops. The two-volume set of papers arising out of Case for Mars III outlined strategies for Mars exploration that touched upon the full spectrum of Mars issues, ranging from mission technical requirements to the public policy and political mobilization needed to turn the vision into a reality.

The Third Case for Mars Conference was a life-transforming experience for me. In the 1980s, after a brief career as a science and math teacher, I had returned to graduate school to study engineering. By 1987, I had obtained Master's degrees in Aeronautics and Astronautics and in Nuclear Engineering from the University of Washington in Seattle, and was working on my Ph.D. in the area of thermonuclear fusion. It seemed to me at that time that the development of fusion power was the most important technological advance that would take place in the last decade of the twentieth century, and I wanted to be part of it.

But by the late 1980s, the U.S. fusion program was in a state of contraction induced by arbitrary budget cuts dictated by OMB director David Stockman and his successors. I am by inclination an inventor, an "alternate concepts" type, and under these conditions the possibilities the fusion program offered for such as me were bleak. Fusion program managers were having a hard enough time accomplishing the central goals of their projects with their contracting funds. Far from initiating new lines of research, the existing set of promising alternate fusion reactor concepts to the mainline Tokamaks—including the Oak Ridge Torsatron, the Livermore Magnetic Mirror, and the Los Alamos Reversed Field Pinch—were all being shut down. I am technically proficient and can do involved calculations when I have to, but the prospect of spending a career working out details within a program of ever narrowing scope did not appeal to me.

Then I went to Boulder, to Case for Mars III. I found the creative atmosphere at the conference wildly exhilarating. The human exploration of Mars was a multidisciplinary problem involving issues of propulsion engineering, orbital mechanics, chemical and biological life-support systems, surface mobility, the search for life, fossils, and other science objectives—local resource utilization, surface power, reconnaissance instrumentation, Martian geography, and human psychology, to name just a few—all in complex interaction. As an inventor with a broad scientific background, being immersed in a problem with so many degrees of freedom was sheer paradise. In exploring the possibilities of this enterprise, I felt myself truly in my element. I liked the people too, being particularly taken with the solidly rational yet visionary McKay and the high-spirited yet tough-minded Stoker.

But there was more to it than that: there was a calling, from across the years, back to the hopes and dreams of my childhood. I had been five when Sputnik flew. I remember seeing a picture of Laika, the space dog, in my *Weekly Reader,* and staring at it for what seemed like hours. To the adult world, Sputnik may have been a terrifying event, but to me, already a youthful reader of science fiction, it was wonderful beyond description, because it meant that the stories of the spacefaring future that I was reading were going to become real. I was nine when Kennedy gave his speech committing America to the Moon; I saw the key part of the speech on TV. As a boy, I had followed every Mercury, Gemini, Ranger, Mariner, and Apollo launch. The human race was moving out, and I wanted to participate either as a rocket designer, an astronaut, or an astronomer, it didn't matter which, the distinctions between these professions being somewhat vague to me at the time. I had learned all the science I could, made sketches of the Moon through my backyard telescope, built rockets and made and tested fuel for them in the basement. I also attempted (unsuccessfully) to build terrariums supporting lizards in a bioregenerative life-support system. In other words, I was with the program, one of Apollo's children. But then, in the early 1970s, America's space effort was wrecked and I had found other things to do.

Now, like the sound of a distant trumpet, the call had been reissued by the Mars Underground. I knew exactly what they were trying to do, and why, and I believed in it strongly. To me, they were a company of heroes, venturing against the odds for the sake of a truly great cause. I decided to join them.

One of the people attending the conference was Ben Clark. The former Viking scientist was now the manager of manned Mars mission studies at the Martin Marietta Astronautics company in Denver. I met Ben, and found him unique. He was a real intellectual who had managed to preserve his brains, his wit, and his spark despite long years of deep immersion in big-corporation culture. We hit it off rather well. I told him I wanted a job working in his group. I had a broad interdisciplinary technical background, exactly what was needed in such studies. He said he would see what he could do.

It took a while, but Ben delivered. Within a year I was working at Martin Marietta Astronautics, doing preliminary design of interplanetary missions.

THE SPACE EXPLORATION INITIATIVE

On July 20, 1989, the twentieth anniversary of the Apollo 11 moon landing, President George Bush got up on the steps of the National Air and Space Museum in Washington, D.C. Flanked by Apollo 11 crew members Buzz Aldrin, Michael Collins, and Neil Armstrong, Bush said that America should commit itself to the goal of sustained human exploration and settlement of space. Then he proclaimed his program: "First, for the coming decade—for the 1990s—Space Station Freedom. . . . And next—for the new century—back to the Moon. . . . And then—a journey into tomorrow—a journey to another planet—a manned mission to Mars."

Thus was born the program that came to be known as the Space Exploration Initiative, or SEI. It was a good start, but it was all downhill from there.

In response to the speech, a sprawling NASA team representing all the centers in the agency, supported by all the major aerospace contractors, went off to figure out how Bush's program could be realized. The team returned three months later with a document entitled "Report of the 90-Day Study on Human Exploration of the Moon and Mars," which soon became known simply as "The 90-Day Report." Before humans could go to Mars, the report said, the nation would need a space infrastructure buildup of thirty years, and the largest and most costly U.S. government program since World War II.

NASA would build the previously envisioned Space Station, but triple its size with the addition of "dual keels" containing large hangars for the construction of interplanetary spaceships. A plethora of additional orbital facilities would be built too: free-flying orbital cryogenic-propellant depots, power-generation stations, checkout docks, crew construction shacks, and so forth. This huge and complex array of facilities would be used to construct and service translunar spaceships (which themselves would require three 80-tonne-lift Shuttle-C vehicles plus one Space Shuttle launch each for their deployment). Those who recalled the single launch required for each Apollo mission scratched their heads and thought, "It wasn't this hard to get to the Moon the last time. . . ." Over the course of a decade, these lunar spaceships would haul to the Moon all the supplies and equipment necessary to build up a massive lunar base complex. Together with the orbital facilities, the lunar base would then provide the basis for building truly huge—1000 tonnes plus—"Battlestar Galactica"–class spaceships for, finally, voyages to Mars. These trans-Mars space cruisers would employ totally new and different propulsion and other technologies than the lunar craft, and thus require vast new development expenditures as well as additional infrastructure beyond those needed to support the lunar missions. Initial Mars missions would require about eighteen months in transit (round trip) with a one-month stay in Mars orbit. As for actually landing on Mars, a small craft would descend to the surface and support a small crew of explorers for two weeks or so, thereby enabling a "flags and footprints" (and little else) human Mars mission to occur. The trans-Mars spaceships would fly out huge and return to Earth orbit tiny, having dropped bits and pieces—fuel tanks, excursion vehicles, aeroshields—in the course of each mission, thereby imposing a massive expense on each "flags and footprints" exercise that followed. The 90-Day Report did not include a published cost estimate, but cost estimates for the program were generated that eventually leaked to the press. The bottom line: $450 billion dollars.

It is doubtful that any kind of program could have survived that price tag. Given its long time lines and limited set of advertised accomplishments on the road to colonizing space, which did little to arouse the enthusiasm of the space-interested public, the 90-Day Report proposal certainly could not. Unless that $450 billion number could be radically reduced, the SEI was as good as dead, a fact made clear in the ensuing months and years as Congress proceeded to zero out every SEI appropriation bill that crossed its desks.

In fact, however, there was no real inner logic underlying the 90-Day Report, or truly new thoughts. Rather, it was a rehash of fixed ideas that harked back to the forty-year-old "Die Marsprojekt," an outline of human missions to Mars that German rocket designer Wernher von Braun and his collaborators had first worked out in the late 1940s and then updated technically to provide the basis for the STG's failed proposal for a humans-to-Mars Apollo follow-on program in 1969. For von Braun and his collaborators, a manned interplanetary mission was part and parcel of a hardware fabricator's wildest dreams: the huge interplanetary spaceship (or better yet, fleets of huge interplanetary spaceships) assembled and launched from Earth-orbiting space stations. What actually occurred on the surface of Mars became an event of secondary interest.

While the STG plan had not been ideal—its use of Opposition trajectories (short Mars stay, long interplanetary coast) on early missions and the building of space stations in geosynchronous and lunar orbits both adding cost without value—the 90-Day Report was much worse. The STG plan had at least been drawn up by people focused on achieving a goal, which was the establishment of permanent lunar and Mars bases using a limited set of modular elements, whereas the 90-Day Report conceived the Mars mission as a kind of Christmas tree on which to hang as many ornaments as possible. Thus the unwieldy 90-Day Report team had proceeded to cast as crucial technologies every existing, planned, or wished-for NASA development program. In order to include everybody in the game, they designed the most complex mission architecture they possibly could—exactly the opposite of the correct way to do engineering.

As a result, the 90-Day Report architecture accomplished about a quarter as much as the STG plan, at twice the cost. Furthermore, this was being proposed at a time when NASA's political and contractor base was much smaller than it had been in the late Apollo period. There were, after all, 500,000 people working for Apollo program contractors in 1969. If the better and cheaper STG plan could not be sold with that kind of built-in political base behind it, the 90-Day Report was clearly hopeless. In point of fact, the 90-Day Report went down like a lead balloon in Congress, and the Bush administration ran for cover, leaving its proudly proclaimed Space Exploration Initiative to linger without funding.

MARS DIRECT

The technical people in my division at the Martin Marietta company all knew what was going to be in the 90-Day Report before it came out. Like many other major aerospace contractors, we had been bound to do detailed designs of the various elements of NASA's mission architecture. I don't know what the folks at the competition thought, but we thought the plan was absurd. Ben and I had a chance to brief upper management at a special strategy retreat held at the ritzy Broadmoor Hotel in Colorado Springs in late 1989, and we made clear to them that unless someone proposed a more practical approach to human Mars exploration, there would be no program. The executives asked us for our recommendation. We requested that Martin Marietta assemble its own "Scenario Development Team" and give it the charter to develop a cost-effective plan for human Mars exploration missions. We asked that we be given a clean sheet of paper for our design, with no requirements input from the marketing department for it to be made pleasing to this or that constituency within NASA. It was that kind of "some pie for everyone" thinking that had made the 90-Day Report such a mess, and it would be hard enough to come up with an attractive human Mars exploration plan without being subject to such imperatives.

Remarkably, the executives agreed. I say remarkably because it is the standard wisdom in the aerospace industry to chime in with whatever "the customer" (NASA or the military) is saying, no matter how flawed it may be. To do otherwise is to risk loss of contracts when the offended government program managers take revenge at those who expressed disrespect for their brilliance. But the sociology of upper management in the Martin Marietta company in the 1989–90 time period was unique. Martin had been the contractor that had actually built and flown the Viking mission for NASA, and the guys who were the hot engineers on Viking in the mid-1970s were vice presidents (and president, and chairman of the board) in 1989. These fellows formed an informal fraternity; they called themselves "Vikings." They wanted to get back to Mars, and they were willing to buck NASA to do it.

So the Scenario Development Team (SDT) was formed, and one of the Vikings, Martin Marietta Astronautics Civil Space vice president Al Schallenmuller was appointed to lead it. Twelve people from across the company were handpicked by Schallenmuller and Ben Clark to comprise the SDT. I was one of them.

Schallenmuller was a remarkable man. Before he came to Martin he had worked in the Lockheed Skunkworks for the legendary Kelly Johnson. Johnson is famous within the aerospace community for the speed with which he developed a string of revolutionary aircraft ranging from the P-80, America's first jet fighter which Johnson developed from program start to first flight in an extraordinary eighty days, to the Mach 4 SR-71 reconnaissance plane. As a result of his time with Johnson, Schallenmuller believed that aerospace programs could and should be done quickly, because that was the way to get the job done and save the taxpayer's money too. He thus felt that humans to Mars within ten years was entirely feasible and appropriate.

Schallenmuller was also a very principled person, with a morality solidly rooted in devout Christianity. On one occasion, he told me that he thought the quest of the Viking biology team to find life on Mars was absurd. There obviously was no life on Mars, he said, and the scientists would have known that "if they had ever bothered to read Genesis." This statement would unquestionably have caused Carl Sagan and many of the other very secular Viking scientists to have kittens. Yet it was Schallenmuller and a bunch of guys like him that got their experiments to Mars. It just goes to show that it takes more than one kind of instrument to make an orchestra.

Schallenmuller felt that for an aerospace contractor to go with the flow and advocate a complex and expensive plan favored by the customer when a simpler and cheaper one was available constituted "waste, fraud, and abuse," and I once saw him lecture a committee of executives from a group of companies to that effect. Thus both by his membership in the Viking fraternity, his background as a Kelly Johnson man, and his morality as a Christian, Schallenmuller was inclined to support plans that sought the quickest, cheapest, and simplest way to get humans to the Red Planet.

The SDT went to work in December 1989. Because there were a lot of very creative people on the team, its members could not agree with one another. So rather than develop one plan, it developed three. Schallenmuller, to his credit, did not try to reconcile the plans to create a consensus party line. This would have been impossible; the plans were based in divergent mission design philosophies and a compromise plan would have degraded all three. Rather, he let each plan follow its own logic—the flowers would bloom where they could.

The first of the plans was that put forth by Ben Clark. He called this plan Concept 6 for a while, but later changed its name to the Straight Arrow Approach, after he realized that a somewhat more evocative name would be advantageous. Basically the Straight Arrow Approach involved a very conventional chemical-propulsion mission design using a true heavy-lift vehicle to minimize the number of launches and a clever self-assembly sequence on orbit to reduce the need for on-orbit infrastructure, while categorically eliminating an array of unneeded technologies that the 90-Day Report had called for. It contrast to the short surface stay, long coast-phase, Opposition Class trajectory adopted by the 90-Day Report, Ben's plan used the much more rational Conjunction Class trajectory, which involves a six-month outbound leg, a year and a half on the surface, and a six-month return leg. The Conjunction Class mission has a round trip time that is 25 percent larger than the Opposition Class, but its useful surface exploration time is greater by a factor of ten and less propellant is needed to drive it. Straight Arrow was thus a good name for this approach; it didn't break much new ground, but it improved greatly on the 90-Day Report through its use of common sense.

The second approach involved maximum use of advanced propulsion technology to reduce mission mass. This one ceased to be a player after mid-1990, so I won't discuss it further here.

The third approach was Mars Direct. This plan was developed by a team led by me and David Baker. Baker was a very sharp engineer with a temperament almost exactly the opposite of mine. I'm a short, mercuric, optimistic romantic; he's a tall, phlegmatic, pessimistic existentialist. My favorite movie is *Casablanca,* his is *Brazil.* This made our collaboration very interesting. But despite our different outlooks, we had come to converging views on the proper design principles for a human Mars mission, and in the early months of 1990 we clicked as a team.

The basic idea behind the Mars Direct mission was this: The reason we want to go to Mars is that it is the planet that has the resources needed to support life. Well, then, why not use those resources to support the mission? In that case, the very same resources that made the Red Planet interesting would also serve to make it attainable.

The idea had solid roots in the history of terrestrial exploration. On Earth, wherever we have explored intelligently, we have always made maximum use of local resources. Thus Lewis and Clark had crossed the American continent with twenty-seven men, hunting and trading as they went. If they had tried to take all the supplies (food, water, and air) needed for their three-year transcontinental trip with them, they would have needed a wagon train of cargo for every man and every horse. Each of the teamsters and horses in those wagon trains would have needed a further wagon train, and the size and scale of the expedition would have exceeded all feasible bounds. Furthermore, even if somehow it had been affordable, such a gargantuan expedition would have been overburdened with its own impedimenta (as the old Roman army used to aptly describe such things), and would have had no mobility to do any real exploration along the way.

Finally, if you ever hoped to settle the areas you are exploring, you would need to learn to use local resources, so it were best to start doing so right from the beginning. The message from the great explorers from Earth's past is very clear: If you want to explore intelligently, travel light and live off the land.

This message was directly germane to the issues Baker and I were wrestling with in the early months of 1990. The key difficulty facing the mission design employed by the 90-Day Report and even the more rational Straight Arrow Approach was the large amount of mass (1000 tonnes for the 90-Day Report, 700 tonnes for the SAA) that needed to be lifted to low Earth orbit to assemble and fuel the huge spaceships employed. These spacecraft needed to be so big because in the mission designs employed, they were required to send a very large payload to Mars. But two-thirds of this trans-Mars injection payload was not necessary equipment—it was simply the fuel and oxygen required to come home. Mars had a carbon dioxide atmosphere, however, and loads of water in its soil, and these are the key ingredients for making rocket propellant. So why not do so, and thereby cut the mass of the mission by a factor of three? If we combined this innovation with splitting the mission into two ships, then each one of these would be smaller by a factor of 6 than that needed in a conventional plan, and thus *small enough to launch to Mars directly* if a booster with the capability of an Apollo era Saturn V (140 tonnes to orbit) were employed.

Thus was born Mars Direct.[17] By combining use of local resources with a split mission strategy and a straightforward Conjunction trajectory flight plan, we would completely eliminate the need for on-orbit assembly of complex futuristic megaspacecraft. Instead, the mission would be done with two direct-throw launches to Mars by a Saturn V or its updated equivalent. Instead of Battlestar Galactica, we had Apollo times two.

Here's how the plan worked: In a given year, call it Year 1, a single booster is used to hurl an unmanned payload massing about 45 metric tonnes onto a trans-Mars trajectory. (Since the booster could lift 140 tonnes to low Earth orbit [LEO], it would be able to send about 45 tonnes on Trans-Mars Injection, or TMI.) The payload of this launch would consist of a number of things, but the primary object involved would be an Earth Return Vehicle, or ERV. This is the craft that will later be used to house a crew of four astronauts on a six-month voyage back from Mars to Earth in the final phase of the mission. In addition to being unmanned at the moment, however, it is also unfueled, but its lower stage tanks contain about 6 tonnes of liquid hydrogen, which will later be used as feedstock for propellant manufacture.

The ERV flies out to Mars on a minimum-energy trajectory, taking eight months to reach the Red Planet, whereupon it uses its aeroshield to generate friction against the Martian atmosphere and capture itself into Mars orbit. Then, after a brief period in Mars orbit to check out systems and verify that surface weather is acceptable for landing, the craft would be redirected back into the Martian atmosphere, this time penetrating much more deeply than before. Plowing through the thicker air, its aeroshell would create sufficient aerodynamic drag to bring the craft to subsonic velocities, after which a parachute would be deployed. As the craft descended slowly on its chute, the aeroshell would be dropped away, and then, at the last minute, the ship would fire its rocket engines to set itself down softly on the Martian landscape.

Once the landing had been accomplished, a light truck running on methane/oxygen engines would be deployed from the ERV. The truck carries a small nuclear reactor with a power of 100 kilowatts (about 130 horsepower) and a windlass containing a large spool of electrical cable. Moving slowly because of the time lag in radio communications between Earth and Mars, the truck would be telerobotically driven a few hundred yards away from the landing site, unwinding the cable off its windlass as it went. Then the reactor would be unloaded, preferably into a crater or a ditch, but the reverse side of a hill would do, anything to put a large chunk of radiation-shielding dirt between the reactor and the main landing area. After this, the reactor would turn on, providing power at the ship.

At this point a small chemical plant aboard the ERV landing stage would go into action, pumping in carbon dioxide from the Martian atmosphere and reacting it with the hydrogen brought from Earth to produce methane and water. This process, known as the Sabatier Reaction, is well-known gaslight-era chemistry, and has been practiced widely on Earth for well over a hundred years. The methane, or natural gas, produced is excellent rocket fuel, and is stored for future use in the ERV's fuel tanks. The water is condensed and then electrolyzed into hydrogen and oxygen. The oxygen is stored; it is the oxidizer needed to make the methane fuel burn in space. The hydrogen is recycled back to the Sabatier Reactor to produce more methane and water. To get the best mixture ratio for burning the methane in oxygen, a second reactor is employed that splits carbon dioxide into carbon monoxide and oxygen. The oxygen produced by this unit is stored, while the carbon monoxide is discarded. In the course of ten months of operation using this rather simple system based on nineteenth-century industrial chemistry, the 6 tonnes of hydrogen brought from Earth would be transformed into 108 tonnes of methane/oxygen bipropellant. Ninety-six tonnes of the propellant produced is needed for use by the ERV to power its return flight. The remaining twelve tonnes is available for use by the expedition's ground vehicles, which all run on methane/oxygen engines.

Launch windows from Earth to Mars occur with a frequency of every twenty-six months, or roughly two years. It will be recalled that the ERV lifted off the homeworld in Year 1, and took eight months to transit to Mars, and then ten months to manufacture its return propellant. So long before the Year 3 launch window opens in month 27, the mission controllers at NASA Johnson Space Center will know that they have a fully fueled Earth Return Vehicle waiting on the Martian surface for the arrival of the crew.

That being the case, in Year 3, two more boosters are launched off the Cape. One sends out another ERV, the other sends a piloted craft with a crew of 4 astronauts in it. Because the crew's return ride is waiting for them on the Martian surface, they don't need to fly to Mars in a gigantic Battlestar Galactica spaceship. Instead a simple habitation ("hab") module will do, something like an 8-meter-diameter tuna can with two decks and a life support system installed. The hab carries provisions for three years, and a pressurized methane/oxygen powered ground rover to allow long-range surface exploration in a shirtsleeve environment.

The hab flies to Mars on a slightly faster trajectory than its accompanying ERV, reaching the Red Planet in six months. On the way to Mars, artificial gravity is provided to the crew by having the hab tether off of the spent upper state of the launch vehicle (which, since it threw the hab on its way to Mars, is coasting outward on the same trajectory too), and spinning the linked assembly to produce centrifugal force.

On arrival at Mars, the tether is cut using a pyro-bolt, and the hab then aerobrakes into Mars orbit. Then, once its system and weather checkout is complete, it is directed to enter the Martian atmosphere and land in the immediate vicinity of the ERV sent out in Year 1.

The surface rendezvous plan has several layers of backup to assure success. In the first place, there is an expert pilot flying the hab, and a landing beacon on the ERV to draw him in, so achieving a close surface rendezvous should be straightforward. In fact, in 1970, Apollo astronauts using primitive guidance and navigation systems and operating without the benefit of a surface beacon managed to land their Lunar Excursion Module within 200 meters of a Surveyor probe that had arrived on the Moon several years before. We should be able to do at least as well today. But in the event we do not, the crew has an alternative. Their pressurized ground rover has a one-way range of 1000 kilometers, which should be more than adequate to deal with the landing errors caused by anyone but a completely incompetent pilot. If that should fail, however, there is a third backup—the second ERV that has been following the crew out to Mars. Even if the crew's landing is inaccurate by distances of planetary scale, the mission can still be saved by bringing the second ERV down to land near them. Finally, if even that should fail, there is a fourth level of backup on the mission, which is that we have the entire crew on Mars where they have natural gravity and substantial radiation protection offered by the Martian environment, and enough supplies with them to last three years. So if worse came to worst, they could just tough it out on the

Martian surface until Year 5, when more supplies and another ERV could be sent out to succor them.

Thus, the mission architecture contains a four-layer defense-in-depth of contingency plans, and each layer results in actually carrying out the mission.

Assuming, however, that the landing actually occurred accurately at site #1, the second ERV would no longer be needed by the crew. It would therefore be landed at a new place, site #2, perhaps several hundred kilometers away. There it would start making propellant that would be used to support the return of the second human Mars crew, which would fly out to it in Year 5, along with a third ERV, which would serve as the second crew's backup but otherwise be used to open up landing site #3.

Thus every two years two boosters would be launched off the Cape. One would send an ERV to Mars to open up a new landing site, the other would send a crew who would explore the most recent previously opened site. Two boosters every two years is an average of one per year. The United States typically launches about six Shuttles per year, which is equivalent to six heavy-lift launches, since the Shuttle has the same takeoff thrust as the Saturn V. If you replace the orbiter with a hydrogen/oxygen upper stage, you can transform it into a booster with similar capability. Thus the Mars Direct plan would require the nation to commit 16 percent of its available heavy-lift capability to support a continuous program of human exploration of the Red Planet. This is something the United States could clearly afford to do.

The first crew would stay on the surface for 1.5 years, taking advantage of the mobility afforded by the high-powered chemically driven ground vehicles to accomplish a great deal of surface exploration. With a 12-tonne surface fuel stockpile, they would have the capability for over 24,000 kilometers worth of traverse before they leave, giving them the kind of mobility necessary to conduct a serious search for evidence of past or present life. This mobility could be extended further simply by attaching a condenser to the tailpipe of the ground rover, allowing most of the water in the exhaust to be recovered and brought back to the base to be reacted with atmospheric carbon dioxide to produce more fuel and oxygen. Alternately, once the crew demonstrates techniques for extracting water from the Martian permafrost, all the fuel and oxygen required for both unlimited rover operations and even the ERV will become producible from local resources, without the need to import any hydrogen feedstock from Earth.

Since no one has been left in orbit, the entire crew will have available to them the natural gravity and protection against cosmic rays and solar radiation afforded by the Martian environment, and thus there will not be the strong driver for a quick return to Earth that plagued all previous Mars mission plans (including von Braun's Marsprojekt, the STG plan, the 90-Day Report, and the Straight Arrow Approach), which were based upon orbiting mother ships with small landing parties. At the conclusion of their stay, the crew would return to Earth in a direct flight from the Martian surface in the ERV, leaving behind their Hab, the reactor, and their vehicles. As the series of missions progressed, a string of small bases would be left behind on the Martian surface, opening up broad stretches of territory to human cognizance.

Such was the Mars Direct plan. In essence, by taking advantage of the most obvious local resource available on Mars—its atmosphere—the plan makes possible a manned Mars mission with what previously had been regarded as a lunar-class transportation system. By eliminating any requirement to introduce a new order of technology and complexity of operations beyond those needed for lunar transportation to accomplish piloted Mars missions, the plan offered the prospect of reducing program costs by an order of magnitude and advancing the schedule for the human exploration of Mars by a generation.

Looked at in the spring of 1990, Mars Direct opened the possibility of humans to Mars by 1999. The plan was intellectual dynamite.

SELLING THE PLAN

Schallenmuller was quite excited about Mars Direct, and he decided to present it, along with the Straight Arrow Approach and other SDT work, to NASA. We gave our first briefing at Marshall Space Flight Center in Huntsville, Alabama, on April 20, 1990. Many of us, including me, anticipated a chilly reception at Marshall for Mars Direct, because the plan was very radical and Marshall was the most conservative of all the NASA centers. We could not have been more wrong. Mars Direct grabbed the attention of the Marshall managers and engineers in attendance in a way that I have never seen happen with any new idea at NASA, before or since. Precisely because they were so conservative in their engineering thinking, they loved it. For the past several months, all the other aerospace companies had been coming in and briefing them on their advanced ideas, which were ever more futuristic and fantastical variations of the Battlestar Galactica concept. We showed them a mission that looked like two Apollo launches. The lights went on in the room. "This is something we could actually do!" one manager said.

Gene Austin, the head of the Marshall group in charge of Space Exploration Initiative work, was so delighted with the plan that he brought Baker and me into his office and sat down with us for two hours discussing the plan and coaching us on how to present it at the various other NASA centers.

So the next week, we went on to NASA Johnson Space Center in Houston, where the briefing also went well. Then it was on to NASA Langley in Norfolk, and NASA Ames Research Center near San Jose. Ames had become the hangout for the Mars Underground: McKay, Stoker, and others had gone there to work after completing graduate school. The reception there for Mars Direct was very warm. Then, at the end of May, I was given the opportunity to deliver the closing plenary at the national conference of the National Space Society. This was the first public presentation of Mars Direct. I received a standing ovation.

A week later, the fourth Case for Mars conference was held in Boulder. Carol Stoker had control of the programming, and she gave Baker and me the bully pulpit of the opening plenary. We basically took the place over. The next day, under the byline of its veteran science reporter David Chandler, the *Boston Globe* ran a front page story headlined "New Mars Plan Proposed." The article was carried by wire and printed in hundreds of other papers as well. Mars Direct was out of the box.

As the summer wore on, Baker and I, singly or together, continued to make presentations at open conferences and NASA briefings. We published a detailed description of the mission as a feature magazine article in *Aerospace America,* the industry monthly, as well. Everywhere we went we made converts, but a counterattack was in the works. Powerful forces within NASA linked to the Space Station program were not happy at all about Mars Direct. Since we didn't use the Station or even its (yet to be created) heritage of on-orbit assembly techniques, we were, in their view, "de-justifying" their program. People in NASA who had been friendly to Mars Direct were told to keep their distance. This slowed us down. Some (but not all) factions in the advanced-propulsion community were also hostile. They felt that Mars Direct was "de-justifying" their programs as well, and argued for mission requirements that only their systems could meet. Refuting the need for these requirements slowed us down some more. What had begun as an intellectual blitzkrieg began turning into trench warfare.

Such an extended fight was incompatible with Baker's temperament. As the difficulty in changing an intellectual paradigm became more and more apparent, and as the NASA bureaucracy's obtuseness in sticking with their $450 billion megafantasy approach continued to result in congressional rejection of SEI funding requests, Baker decided the effort was futile. In February 1991, he quit Martin to go back to school at the University of Colorado for his Master's degree and to start his own consulting firm.

But I kept going, writing papers that dealt with various secondary issues, briefing one person or group after another. In the spring of 1992, I got a chance to brief Mike Griffin, the new NASA Associate Administrator for

Exploration, and he immediately became a strong supporter of Mars Direct. That summer, Griffin briefed incoming NASA Administrator Dan Goldin, and he also became supportive. Then, in the fall, Griffin had me go back to Johnson Space Center to present a detailed set of briefings on the plan. He would see to it that they took it seriously.

The exploration group at JSC certainly did take it seriously. We had an extended set of briefings, and when we were done they expressed agreement with the fundamental set of principles underlying Mars Direct: (1) direct launch to Mars, (2) no on-orbit assembly, (3) use of Martian resources starting on the very first mission, and (4) long-duration stays (i.e., Conjunction trajectories employed) starting on the very first mission. But they still had concerns. In particular they felt the Mars Direct ERV was too small.

So JSC's lead mission architect, David Weaver, and I got out the chalk and worked out a compromise plan, which I call Mars Semi Direct. The Semi Direct plan uses three launches per mission. One launch sends a Mars Ascent Vehicle (MAV), which makes its propellant on the Martian surface as in Mars Direct, but only enough to reach a highly elliptical (loosely bound) orbit around Mars, where it performs a rendezvous with an Earth Return Vehicle delivered to that orbit from Earth by a second booster launch. The third booster flies the crew out to Mars in a Hab module very similar to that used in the Mar Direct plan, where it performs a surface rendezvous with the MAV. Because the plan employed three launches instead of two, more cargo could be delivered and the crew size was increased from four to six.

The Johnson Space Center exploration group embraced the Semi Direct plan, calling it the Design Reference Mission. They then had their costing group—the same folks who had generated the $450 billion cost estimate for the 90-Day Report—estimate the price of a human Mars exploration program based on this mission architecture. Their answer: $55 billion.

I protested that they had included too much or too large equipment of various sorts. For example, instead of the Mars Direct ground rover, which was about the size of a small 4-by-4, theirs looked like a school bus. Furthermore, I argued that they really didn't need six people in the crew, four would do, and reducing the crew logistic demands would shrink the size and cost of the required booster. If they made the mission leaner, I said, program costs could be reduced closer to the $30 billion level.

They answered that these points were essentially irrelevant. It didn't really matter that much whether the cost estimate was $30 billion or $50 billion. The real issue was that the Design Reference Mission (DRM) had brought the estimated cost of human Mars exploration down from hundreds of billions of dollars to tens of billions. Tens of billions is what the United States routinely spends on any number of medium-sized military procurement programs. The V-22 Tilt Rotor airplane program, for example, cost $40 billion. A human Mars exploration program costing hundreds of billions would never fly. One costing tens of billions was possible in principle. That was the key thing.

They were right. In the summer of 1994, word about the Design Reference Mission leaked out. *Newsweek* science reporter Sharon Begley made it the centerpiece of her July 1994 cover story timed to coincide with the twenty-fifth anniversary of the Apollo 11 Moon landing. "A manned mission to Mars?" *Newsweek* asked. "The technology is already in place. And at $50 billion—one tenth of previous estimates—it's a bargain."

NASA's 1993 adoption of the low-cost Design Reference Mission came about two years too late to save the SEI. In November 1992, Clinton-Gore replaced Bush-Quayle in the White House, and they rapidly dismantled what was left of the initiative. Thus, by the time NASA had adopted a program that could be sold to Congress, there was no longer an administration interested in trying to sell it.

Nevertheless, the conversion of the NASA mission planners to the Mars Direct paradigm was a major victory, because it meant that the next time a president went to NASA and asked how much and how long a humans-to-Mars program would require, he would not get back the politically impossible answer of thirty years and $400 billion. Instead the reply would be something like ten years and $50 billion. This would at least give the program a fighting chance.

Furthermore, while the Clinton-Gore administration may not have been interested in hearing about how humans-to-Mars could be made possible, there were others who were. Immediately after the *Newsweek* article was published, I began to receive invitations to brief some well-placed people, such as soon-to-be Speaker of the House Newt Gingrich. I also received a phone call at my desk from a woman named Laurie Fox.

BIRTH OF A BOOK

Laurie was a literary agent, and she was very taken with the *Newsweek* article. "Do you realize you have a book here?" she said. "Would you like to write a book?" she asked. I said that I would love to, but that I had once written a spy novel and gotten nowhere finding a publisher. "Oh," she said. "Have you ever been a spy?" No. "Did you have a literary agent representing you?" No. "Well, that explains it. Of course no one would read your manuscript. This will be different. Stick with me, I'll get you published."

She did. She showed me how to write a book proposal and encouraged me to get together with my old friend former *Ad Astra* magazine editor Richard Wagner, so we could field the kind of engineer/wordsmith writing team that publishers find attractive. Then she energetically circulated our proposal to forty-three publishing houses, obtaining forty rejections from publishers who could not imagine how a book on human Mars exploration could interest anyone, and three offers. The offer I liked best came from the Free Press, a subdivision of Simon & Schuster.

The Free Press was headed at that time by Adam Bellow, the son of author Saul Bellow, and included in the ranks of its editors a man named Mitch Horowitz. Mitch was an intellectual who loved books about ideas and bold adventures, and Mars exploration appealed to him on both counts. It was Mitch who had convinced the Free Press to make an offer for the book, and who would be its editor if the manuscript was sold to that house. A good relationship between editor and author is essential for a book project to prosper, and in talking with Mitch I decided that I liked him better than the alternatives at the two other houses making offers. So the deal was struck.

The choice proved wise. The original proposal for *The Case for Mars* included a gamut of things, including a guided tour of Mars and a discussion of the folklore of Mars, that we had thought might make the book popular. Mitch made us take all that stuff out. It's been done before, he explained, by people like John Noble Wilford (in *Mars Beckons*). What people want from you is the hard stuff, the plan, a real how-to book explaining the way to get humans to Mars. So you need to keep it tight.

This was fine by me. So we went to work. We started with Mars Direct, and explained how the plan worked. Then we explained the thought process that went into conceiving it, and refuted the objections to its feasibility. Then it was on to the future; base-building, settling Mars, and terraforming—laying out a summary how-to manual for creating the first extraterrestrial civilization.

Mitch required a substantial rewrite of our first draft, but accepted the second. He then entrusted the manuscript to the Free Press's copy editor, who returned it to us covered in handwritten notes that represented requests for a host of minor alterations. Fortunately, as a former editor, Richard was able to decipher these markings, and he turned it around without delay. The book then went to the Free Press's production department, and was scheduled for publication in December 1996.

But then, in August 1996, the Alan Hills meteorite team published its findings, and every news media outlet in the world went abuzz with talk about Mars exploration. Mitch convinced the Free Press to "crash" the book's production schedule, accelerating release from December to September.

The wisdom of this was demonstrated instantly. Within three weeks of its September publication, the entire first hardcover printing of *The Case for Mars* had sold out. Mitch got the Free Press to do a second hardcover printing, and when that sold out a third, and then a fourth. The reviews were great too. Writing in the *New York Times Book Review,* Dennis Overbye called the book "one of the most provocative and hopeful documents I have read about the space program in 20 years." Simon & Schuster decided to issue a paperback edition, and foreign rights were sold. Soon *The Case for Mars* was hitting bookstores across the globe in German, Japanese, Polish, Chinese, and eventually Russian.

The net result of all of this was that I started getting mail. Lots of mail. In fact I received over 4,000 letters and e-mails from around the globe. The letters came from bankers in Paris, engineers at JPL, opera directors in New York City, astronauts in Texas, twelve-year-old students from Poland, war hero widows from Virginia, and firemen from Saskatoon. They asked me about propulsion, life support, terraforming techniques, power production, politics, and property rights. You name it, they asked it. But fundamentally, beneath all the varied inquiries there was one basic question: How do we make this happen?

I looked at this incredible mass of correspondence and I thought to myself, If only there were a way to bring these extraordinary people together, their combined abilities, talents, and connections could well be enough to make humans-to-Mars a reality.

Then, on July 4, 1997, NASA landed its Pathfinder probe on Mars. Within twenty-four hours, the NASA Mars website received 100 million hits, causing overload and failure of large parts of the Internet. One hundred million is more than the number of people who vote in the United States. One hundred million people times one hundred dollars is $10 billion, a sum probably sufficient to send humans to Mars if the program were conducted in the private world.

THE FOUNDING OF THE MARS SOCIETY

It became abundantly clear that those people who believe it is important for a positive future for humanity to expand to Mars had it in their power to achieve that result, if only they could be mobilized. However, while the harvest was plentiful, the gatherers were few. Organization was needed.

I phoned Chris McKay and broached the idea of setting up a Mars Society. He agreed the time was right. I had left Lockheed Martin (as Martin Marietta was renamed after the 1995 merger) in early 1996 to found my own company, Pioneer Astronautics, and therefore had the independence required to lead such a project.

As originally conceived, the Mars Society would have three primary modes of activity. First, we would engage in broad public outreach to spread the vision of Mars exploration. Second, we would attempt to educate the political class to embrace the goal of human Mars exploration, and we would mobilize the public to defend both robotic and human Mars exploration programs around the world. Third, we would attempt to initiate Mars exploration on a private basis, starting with small projects and using the credibility earned from their success to mobilize the larger technical and financial resources needed to accomplish ever bigger ones.

So in the late fall of 1997 the banner was raised. First to rally to the standard was my wife Maggie, followed soon by Carol Stoker, Tom Meyer, and most of the old Mars Underground. We formed a Steering Committee, and more people joined, including Richard Wagner, former NASA Associate Administrator for Exploration Mike Griffin, well-known space law attorney Declan O'Donnell, astronaut Scott Horowitz, and noted science-fiction authors Kim Stanley Robinson and Greg Benford. We took the addresses from the 4,000 letters I had received from *The Case for Mars* correspondence and turned them into a mailing list, sending out information bulletins and a mailing calling for a founding convention to be held in August 1998 in Boulder.

The organization became active in May 1998, and almost immediately scored a significant success in June by launching an e-mail barrage to the Senate Appropriations Committee, causing it to restore $20 million (of $60 million that had been cut) to the 2001 Mars mission. This was enough to get the small Marie Curie rover back on the mission, replacing the larger Athena rover unit that had been pulled. The material consequences of this victory ultimately proved to be temporary in nature, as the whole 2001 lander mission was canceled by NASA in the wake of the

1999 Mars probe failures. But the effect of the mobilization on the newborn Society's morale was permanent—people learned their actions could have an effect. This lesson in political activism was to pay off later, when a similar mobilization of the Society was successful in preventing a German pullout that would have scuttled Europe's Mars Express mission.

Then, in August 1998, we held our Founding Convention in Boulder, Colorado.

The old Mars Underground "Case for Mars" conferences had peaked in 1987 and 1990, but after the collapse of the Space Exploration Initiative made the cause seem lost, had gone downhill fast. In 1993, attendance had dropped to 250, falling further to 120 in 1996, prompting science reporter and longtime Mars Underground observer Oliver Morton to write a column in the *Economist* entitled "Case Closed." However, our preorganizing for the Mars Society Founding Convention had set off a kind of buzz throughout the Mars-interested community. There was a sense that something exciting was afoot. At the end of April there were only 46 registrations, and just 54 more came during May. But in June, things broke loose with 227 more registrations coming in one right after another over the fax machine, and then 300 more during July. The press got wind of it; it was going to be "The Woodstock of Mars." I started getting phone calls from the *New York Times,* the *Washington Post,* and dozens of other media asking for information and requesting press credentials.

The evening before the convention was to begin, we held preregistration in the University of Colorado planetarium. The place was mobbed, and we had no staff. So led by Lorraine Bell and Paulette Gerardy, a group of the attendees volunteered themselves as staff and took over the registration process.

The next morning, Thursday, August 13, the convention proper began in CU's Glenn Miller Ballroom. The place was crackling with excitement. There were 750 people there, from dozens of countries. Every NASA center, every national lab, and nearly all the top universities in the United States were represented. But they were not only there from North America and virtually every country in Europe, they were there from Japan, China,

Australia, New Zealand, Argentina, Israel, Egypt, and even Mozambique. The international press was there, in force, interviewing everyone. Only about a quarter of the people there were from the old Mars Underground. The rest were new people, meeting one another for the first time. There were Green Berets and folksingers, astronauts and artists, Lutheran ministers and science-fiction authors, businessmen, arctic explorers, plenty of scientists and engineers, and a spirited contingent of Air Force Academy Cadets.

I got up to open the convention, beginning, "Hello, I'm Robert Zubrin, president of the Mars Society." Then, overwhelmed by the sense of the crowd, I burst out, "and boy am I glad to see you!"

It turned out to be exactly the right thing to say. The hall broke out into warm laughter, and the crowd of strangers turned into friends. I proceeded to give the best speech of my life. I briefed them on Mars Direct, and the wide-open future to which it could lead, and how we could make it happen. I concluded quoting Pericles' speech in praise of Athens; "Future ages will wonder at us, even as the present age wonders at us now." The audience saw what I meant. They stood up and cheered. The conference turned into magic.

We had one great speaker after another. Everett Gibson of the Alan Hills Meteorite team presented a vigorous defense of the team's biologic interpretation of the stone's more intriguing features. Pascal Lee, then of NASA Ames Research Center, gave an overview of recent research in the Haughton Impact Crater on Devon Island in the Canadian Arctic. Rob Manning of the Jet Propulsion Lab presented a preview of NASA's current robotic Mars exploration program. Jacques Blamont of the French space agency CNES, the world's leading advocate of balloon-borne Mars exploration, showed the audience how the Red Planet could be explored from the air.

That was just the agenda for Thursday morning. On other mornings, we had plenaries given by Chris McKay on the Search for Life, Carol Stoker and Larry Lemke on Mars Airplanes, former NASA Associate Administrator for Exploration Mike Griffin on the politics needed for a successful Mars program, and Gemini-Apollo-Shuttle veteran John Young giving the astronaut's point of view.

But the plenaries were only for morning starters. During the afternoons, we had five simultaneous speaking tracks running, allowing the attendees to present their own material. Some 180 talks were given, covering everything from the technical aspects of Mars exploration to the social, financial, political, philosophical, and even artistic. These talks were a joy to behold, as scores of people joined in the fun of the creative play of ideas involved in opening a new planet to humanity. Particularly astonishing was a girl named Kathleen Bohne, whose talk titled "Our Future on Mars from My Perspective as a 12 Year Old" "blew everyone away" (as one audience member remarked) by her moving honesty and vision.

There was controversy too. On Friday night, we had a panel dealing with the ethics of terraforming. Lowell Wood, a nuclear weapons scientist from Lawrence Livermore National Lab, used the term "Manifest Destiny" to describe humanity's obligation to bring dead Mars to life. This caused quite a ruckus among some of the more "politically correct" members of the audience. Terraforming is not Manifest Destiny marching westward, they said, it's more like gardening. It should not be thought of as an assault on nature but as an action giving birth to nature. It's not war, it's love.

Well, so it is. The debate caused tempers to flare, but I liked it. Despite wild disagreement over terminology, it showed that the vision of humanity settling Mars had power across the political spectrum. We had right-wing bomb-makers and left-wing feminists and antiwar activists in the same room, all joining in the same enterprise. The cause of Mars will need both pilots and poets. The political diversity exemplified at the terraforming debate was exactly what the Mars Society had to have if it were to have any chance of moving the nation.

On Saturday afternoon, the Steering Committee met for the first time. We resolved on a political program, including doubling the budget of the robotic Mars exploration program with much of the added funds going to start a new "Mars Discovery" program in which Principal Investigator–led teams from academia, industry or various NASA centers would compete their creative Mars mission ideas for funding. (Both these objectives have since been achieved. The Mars Society's Mars Discovery concept was implemented by NASA in 2000 under the title Mars Scout. The first such mission is scheduled to fly in 2007.) We also resolved on a campaign to have congress allocate 1 percent of NASA's budget, or $140 million per year, to start developing technologies to enable human exploration of Mars. (This has not yet been achieved, but in 2002, Congressman Nick Lampson, a Democrat from Texas, introduced a bill, the Space Exploration Act of 2002, that would have provided almost that much.)

The most exciting resolution adopted at the Steering Committee meeting was the decision to launch our first project. We decided that the Mars Society would commit itself to building a simulated human Mars exploration base on Devon Island in the Canadian Arctic. Located at 75 degrees latitude, just 1500 kilometers from the North Pole, Devon was site of a huge meteor impact 23 million years ago. In the context of the local polar desert climatic conditions, this has resulted in a set of geologic features that are believed to resemble those that may exist on Mars. For this reason, NASA scientists started exploring this area in 1997 in order learn about Mars by geologic analogy. On the suggestion of Pascal Lee, one of the scientists involved in this work, we decided to build a base, including a prototype Mars habitation module equipped with suitable field lab instrumentation in the area. This base could be used to support the ongoing field work, as well as a test bed for Mars life-support equipment, field mobility vehicles, permafrost penetrating drill rigs, astronaut training, testing out operational procedures and field exploration strategies, and many other invaluable purposes.

The convention came to a head at the Saturday night banquet. I announced the results of the Steering Committee meeting. Then I read out the Founding Declaration of the Mars Society. This stirring document had been drawn up by science-fiction writer Kim Stanley Robinson, Chris McKay, Richard Wagner, and I, and an oversized version of it had been posted in the lobby outside the main ballroom for signing since Thursday morning. By this time, almost everyone had signed it, but I now read it out for a vote. "The time has come for humanity to journey to Mars. . . . We must go for the knowledge of Mars . . . for the knowledge of Earth . . . for the challenge . . . for the youth . . . for the opportunity . . . for our humanity . . . for the future."

The vote was unanimous. The Society's principles thus adopted, we opened up two microphones for members to stand up and speak out in open town-meeting style as to what we should do next. At 11 P.M., I threw a chart on the overhead projector depicting a map of the ballroom. "If you are from France, go there . . . Ohio, there . . . California, there . . . Australia, there . . . Okay, folks, form your chapters."

Twenty-five Mars Society chapters were formed that evening. When midnight came, and the university staff threw us out of the ballroom, everyone moved out onto the patio, where the discussions continued far into the night.

The Mars Society had been born.

4. DEVON ISLAND

The continents have moved a bit since then, but 23 million years ago Devon Island was already in the far north. The early Miocene was a warm era, however, so despite its polar location, Devon's general environment was comparable to what we now see in the subarctic. Forests of small conifer trees covered its landscape, offering ample homes for groups of giant rabbits and herds of tiny rhinoceri.

Suddenly, disaster struck. From the depths of space came a messenger of destruction. A small asteroid, perhaps a kilometer in diameter, weighing several billion tonnes and moving at a velocity of perhaps 20 or 30 kilometers per second, smashed into the island. The impact released energy comparable to that of 10 million Hiroshima-sized atomic bombs. In a blinding flash, everything within hundreds of miles of ground zero was killed nearly instantly, and if present-day "nuclear winter" theorists are even within a factor of a thousand of being right, a cloud of dust and smoke must have been thrown up sufficient to cause severe global cooling for several years.

When the smoke cleared on the island, a crater remained, some 20 kilometers in diameter and surrounded by a bizarre landscape of ridges made of the condensed powder of vaporized rock and shock-fractured stones whose intense flash heating had denuded them of all nitrates and other organic content.

Millions of years passed. The basin of the crater became a lake. Erosion modified the walls of the crater somewhat, but the structure was so huge that not even the repeated advance and retreat of the Ice Age glaciers could destroy its fundamental shape. In time, the island became visible to the eyes of humans.

The first to see it were the paleoeskimos, primitive Pre-Dorset and Dorset people, wandering eastward across the high Arctic from their Alaskan homeland toward their Greenlandic destiny. Then, about a thousand years ago, the Thule people arrived. Ancestors of the Inuit of today, the whale-hunting Thule brought with them a complex social structure based on the leadership of the whaleboat captain. Possessed of an advanced Neolithic technology, the Thule had large skin-covered boats and used sleds pulled by dog teams. They displaced the peaceful and shy Dorset (who are remembered in Inuit folklore today as the Tuniit or Tunirjuat) and established a culture that spanned most of the North American Arctic.[18] In some places they pushed south, becoming caribou hunters, until their expansion was contained by the Cree and other Canadian Indian tribes, who became their perennial enemies. Remains of Thule whalebone-framed dwellings can still be seen in Resolute Bay on Cornwallis Island, just across a narrow strait from Devon.

With the arrival of Martin Frobisher on Baffin Island in 1576, the Inuit and the high Arctic entered European history. Europeans saw the easternmost coast of Devon Island for the first time through the eyes of Henry Hudson's former first mate Robert Bylot and his pilot William Baffin in 1616, who identified the entrances to the Jones and Lancaster Sounds that run along the island's northern and southern coasts. But no European was to see more of Devon until 1819. In that year, British Navy Lieutenant William Edward Parry commanding the sturdy bomb ship *Hecla* and accompanied by Lieutenant Mathew Liddon in the gun brig *Griper,* passed the entire southern coast of the island on his right as he sailed west through the entirety of Lancaster Sound into the waters beyond. Parry was on a voyage in search of the fabled Northwest Passage from the Atlantic to the Pacific. In an amazing sprint, he actually made it to Melville Island, about 112 degrees west, until ice blocking his way forced him to winter over and then turn back. As documented by Pierre Berton in his magnificent book *The Arctic Grail,*[19] from which the following account of nineteenth- and early twentieth-century Arctic exploration is drawn, no other European vessel was to penetrate as far into the Arctic for another thirty years, and no sailing ship entering from the east ever went farther.

But more than one tried.

Sail, sail, adventurous barks! Go fearless forth,
Storm on his glacier-seat, the Misty north—
Give to Mankind the inhospitable zone
And Britain's trident plant in seas unknown . . .

—Eleanor Anne Porden, 1818

Jane Griffin was a remarkable woman. An exquisite beauty, she had dozens of suitors, all of whom she spurned. Dandies and talented men alike threw themselves at her feet, but to no avail. Among those left pining at the base of her pedestal was Peter Mark Roget, who, sublimating her rejection, went on to create Roget's Thesaurus. But while Jane did not fail to make her impression on the social scene, she had a deeper interest—the world of ideas.

Miss Griffin was a formidable intellectual. Born in England in 1792, she had, from the age of nineteen, dedicated her life to the systematic improvement of her mind. She had a plan for organizing her time and read a hundred books a year, but only serious nonfiction works on education, religion, foreign countries, and social problems. Novels were not for her. She was interested in serious subjects, like reforming prisons and improving

Britain's schools. Her best friend was an equally brilliant woman, the poetess Eleanor Porden. Porden's poems are forgotten today, but as the author of a 16-canto epic about Richard the Lionheart called *Coeur de Lion,* she was widely regarded as one of Britain's leading literary lights. She was also a wit, and despite a frail constitution, had a gay nature that attracted men across continental distances. A devout yet lighthearted Anglican, she was a fierce patriot who firmly believed in Britain's civilizing mission in the world.

Captain John Franklin was an Arctic explorer and a heroic veteran of the decisive naval battle of Trafalgar. In 1818, he made an attempt at the North Pole, which did not get very far. But on his return to England, he was greeted by a newspaper carrying poetry by Eleanor Porden, lauding his intrepid British spirit. He decided to look her up.

The muse and the explorer took to each other instantly. On August 19, 1823, after a four-year courtship interrupted by another Arctic expedition, they were married.

Their life together did not last long. Less than two years after the wedding, Eleanor was dead of tuberculosis. She had, however, lived long enough with Franklin to be able to pass some positive reviews on to her friend. Jane decided to make the sea captain her project. In December 1828, at the incredible age of thirty-six, she married him.

In many ways, Jane was a much better match for Franklin than Eleanor. In contrast to the delicate poetess, Jane Griffin was tough and athletic. When her husband was posted to the Mediterranean fleet, she went with him, and spent her time hiking long distances to visit ancient ruins in Greece and Egypt. Together with her explorer husband, she climbed mountains. For a man like Franklin to have found a woman in that day and age who could share such pursuits must have been sheer joy, and he loved her deeply. She loved him in return, and wanted to travel the world with him. When he was appointed to be Governor of Tasmania in 1836, she went there too.

Jane, now Lady Franklin, was, however, a woman of strong liberal convictions, and as such she was deeply offended by the hideous conditions of the prison colony. She created a commission to protect the women convicts. She tried to intervene on behalf of the aborigines. She started a college and a scientific society. She had her husband reverse the unjust dismissal of a doctor who was trying to help the poor.

The colonial pooh-bahs were outraged. They ran the local newspapers, and used them to mock Franklin as "a man in petticoats" for supporting his wife in such effrontery. The *Colonial Times* minced no words. "If ladies will mix in politics, they throw from themselves the mantle of protection which as females they are fully entitled to. Can any person doubt that Lady Franklin has cast away that shield—can anyone for a moment believe that she and her clique do not reign paramount here?"

Governor Sir John Franklin was not a man to put up with this kind of stuff. So he suspended the reactionaries' ringleader, Colonial Secretary John Montagu, and sent him back to England. Unfortunately, once there, Montagu went to Secretary of State for the Colonies, Lord Stanley, and had himself reinstated and Franklin dismissed instead.

Franklin was publicly humiliated and insulted in the Tasmanian press, which published word of his firing under the headline "Glorious News!"

On January 12, 1844, with the mocking laughter of the Colony's entire upper class ringing in their ears, Sir John and Lady Jane Franklin went down to the pier to board ship for England. They found the docks lined with people.

Two thousand of the colony's poorer folk had turned out to cheer for them.

The shouted thanks and farewell wishes of Tasmania's lower orders may have warmed Franklin's heart for a moment, but deep inside he felt disgraced. His reputation was in ruins. He had to restore it, and the only way he could do that was to go back to where he had established it in the first place. A simple, direct man, Franklin stood no chance against the Montagus in the Byzantine world of colonial bureaucratic politics. The cold, clear Arctic was his true element, where man tested his mettle against nature in honest, open combat. He was almost sixty years old. His last Arctic adventure had been nearly two decades in the past. But there was no other way.

Franklin would go back to the Arctic. He would succeed where Parry had failed. He would find the Northwest Passage, or die trying.

Nearly sixty, Franklin was by most accounts too old for such an expedition. He was also growing fat. Many Arctic expeditions involved a lot of trekking and heavy hauling. They all involved cold weather. How would the aging and out-of-shape Franklin cope? There was no problem, Franklin replied. He would be safe, snug, and relaxed aboard ship. Lord Haddington, the First Lord of the Admiralty, was doubtful. But knowing Franklin's distress, the old Arctic hands in the Navy rallied to his cause. "If you don't let him go, the man will die of disappointment," Parry told the Admiralty. The mission was approved.

It was a very bad decision. Franklin may have been a kind and lovable man (as British naval officers went), but he represented the Royal Navy's most backward ideas on how to conduct Arctic exploration. For reasons of pride and prestige, he decided to go with a big expedition. This was a mistake; large-scale expeditions of exploration into the wilderness are an invitation to disaster, because it is almost impossible to feed them off of local resources. Franklin guaranteed that this would be the case by not taking any big-game hunting rifles or anyone skilled in big-game hunting. Instead, all supplies for a three-year voyage were taken from his port of departure. This put the expedition on a time fuse for famine. It was worse than that, however, because Franklin was relying on his lemon-juice supply for protection against scurvy. This could last three years if rationed in the standard dose of one ounce per sailor per day. Such doses were sufficient to ward off scurvy on ordinary naval voyages of three to six months, but were entirely inadequate for a three-year voyage. If Franklin's men had relied on hunting, the fresh meat would have protected them against scurvy. This is how the Inuit survive in the Arctic. But Franklin refused to learn from the Inuit. He even refused to learn from the British Navy, whose more experienced hands by this time knew that small shallow-draft vessels were necessary to make progress through the Arctic archipelago.

Franklin would have none of that. He would travel in style on the two large modern steam frigates *Erebus* and *Terror,* each with a draft more than twice as great as previous British Arctic exploration ships that had been abandoned after running aground.

On May 19, 1845, Franklin set sail. He had at his command 134 men aboard two fine ships. The vessels were well loaded with expensive Victorian china, cut glass, mahogany desks, splendid dress uniforms, and excellent silverware, but there was not a hunting rifle, a sled dog, or a harpoon between them. On July 26, 1845, they were spotted by a Scottish whaler off Baffin Bay. They were never seen again.

Two years passed with no word from Franklin. The Admiralty counseled patience, but by November 1847 Jane could stand it no longer. She called a meeting in her home to plan a rescue expedition. They ended up planning three. Two ships would enter the Arctic from the east, and two more would travel around Cape Horn to enter from the west. A third, under Franklin's friend Sir John Richardson, would sail to the Canadian northwest and then trek to the Arctic overland. Jane wanted to join this overland expedition herself, but was eventually convinced she could do more good by staying behind in England and using her influence to mobilize additional search parties.

Mobilize she did. She prodded the British navy into launching a series of rescue attempts. She outfitted five ships at her own personal expense and sent them. When her money ran out, she raised more funds by public subscription (The people of Tasmania donated 1,700 pounds in response). She wrote to President Zachary Taylor and had the American government launch expeditions, and traveled to the United States to convince American millionaire Henry Grinnell (after whom Devon Island's Grinnell Peninsula is named) to fund several more rescue ships privately. She wrote to Czar Nicholas, and despite the Crimean War, got the Russians to aid in the search. She simply would not give up, and her intrepid spirit made her a heroine, not only to the British public but to explorers worldwide as well.

As to a latter-day Joan of Arc, brave men everywhere rallied to her call. An entire generation of Arctic explorers was summoned into existence, including men like the American Elisha Kent Kane, the memoirs of whose exploits in the Franklin search were to inspire Robert Peary and Roald Amundsen to take up the Arctic quest half a century later. In all, over the next twelve years, Lady Jane Franklin was to mobilize some fifty expeditions to try to rescue her lost husband.

One of the men Jane recruited to the cause was William Penny. Penny was a whaler who had worked in the Arctic since he was eleven, with sixteen years' experience commanding whaling vessels. He volunteered to serve Jane without fee, and in March 1850 sailed north in command of the *Lady Franklin* and the *Sophia* (named after Jane's niece and confidante).

On August 27, 1850, Penny had his ships anchored off Beechy Island, actually a peninsula connected at low tide to the southwest edge of Devon Island. By coincidence, he was joined there by the *Advance,* an American vessel funded by Grinnell and carrying Elisha Kent Kane as ship's surgeon, and the *Felix,* a British ship privately funded and commanded by seventy-three-year-old veteran explorer John Ross. Suddenly some of Penny's sailors came running toward the ships over the ice. "Graves, Captain Penny!" they shouted. "Franklin's winter quarters."

There were indeed graves, but only three. Franklin had clearly stopped at Beechy to bury three men. He had also placed a cairn nearby, but there was no message in it. The question was, Which way did he go from there?

Penny decided to sail to the north. He took his ship up through the Wellington Channel, which separated Devon Island from Cornwallis Island. On Devon's western shore he found a piece of British elm timber. It was a clue. Perhaps Franklin was on Devon Island.

Penny had literally grown up in the Arctic, and knew much more about it than the British or American naval officers. He had therefore done something that they had not. He had brought dogs, and a dogsled, and an experienced Greenlandic Danish dogsled driver named Carl Petersen. Penny and Petersen took the dogsled and ventured inland on Devon Island. They did not find Franklin, but they became the first Europeans to explore the island's interior. They were also the first European explorers to demonstrate the effectiveness of Inuit-style dogsled travel in the Canadian Arctic archipelago—the technique that ultimately was to prove Amundsen's salvation when he finally succeeded in winning through the Northwest Passage in 1903–5.

Franklin's fate was finally discovered by Francis M'Clintock, commanding the Fox, in an expedition financed by Jane Franklin in 1859. In a cairn on Victory Point on the northwest coast of King William Island, M'Clintock found a document written by Franklin's Lieutenants Crozier and Fitzjames dated April 27, 1848, that told the whole story. Sir John Franklin had headed north through the Wellington channel separating Devon Island from Cornwallis Island, just as Penny had suspected. Instead of heading inland on Devon, however, he had circumnavigated Cornwallis Island and headed south, to die near King William Island on June 11, 1847. Many of the younger men had lasted longer, but after nineteen months of being trapped in the ice, they had decided to abandon their ships and attempt a march to the south.

It was later determined that those attempting the southward trek dropped in their tracks manhauling sledges filled with the expedition's valuables, with the large majority felled by scurvy.

Franklin's men need not have died. Half a century later, Amundsen and his six-man expedition were also frozen in on King William Island for two years. But Amundsen had learned from Penny. Traveling by dogsled, his men were able to hunt the highly mobile herds of caribou that abound in the area and, quite literally, lived off the fat of the land until the ice melted and their little sealing boat, the *Gjoa,* broke free to sail them through to Alaska in triumph.

Lady Jane Franklin's quest to rescue her husband and his men thus ended in failure. But the search was not without positive results. In the course of some fifty expeditions inspired by the Franklin cause, the Canadian Arctic was mapped.

COOK'S WINTER ON DEVON

Years passed. Then, in 1908, the American Frederick Cook made his attempt for the North Pole.[19] Basing himself on the Amundsen small-group model of live-off-the-land exploration, Cook decided that his party would include only three people: himself and two young Inuit named Ahwelahtea and Etukishook (Cook's spelling).

Cook went first to Anoatok, in Greenland, where he joined his Inuit companions and had himself fitted out by natives in Inuit furs. Then, traveling by dogsled, the three crossed the Kane Basin and Ellesmere Island to reach the northern tip of Axel Heiberg Island on March 8, 1908. According to Cook, he then made his dash for the Pole, some 600 miles to the North, crossing the pack ice to reach it on April 21. This claim is controversial to this day, being disputed then and since by partisans of Cook's rival Robert Peary. What happened on Cook's return, however, was much more remarkable.

From however far north he actually managed to get, Cook returned south, passing to the west of Axel Heiberg. Lost in a gray mist, the group's food ran low, and the men and dogs together had their food ration reduced to 6 pounds of meat per day. That was not enough to live on, so the men started to eat the dogs.

Eventually, they reached the Penny Strait, between Bathhurst Island and Devon Island. Since it was known from Penny's exploration that there was game on the north shore of Devon Island, they turned east at that point, and crossed the Grinnell Peninsula, heading in the direction of Bear Bay and Jones Sound. This route took them within a few miles of Devon's giant impact crater, and possibly right along its edge.

They pushed on, finally reaching Cape Sparbo, on Devon's north coast, in September. It was getting cold and dark, and they had no choice but to stop and winter over.

Their condition was desperate. They were out of food, their dogs were gone, and they had only four rounds of ammunition among them. They made their home in a cave, only to discover from a skull inside that they were living in a prehistoric tomb. They had no ammunition left to hunt with, so they made bows. These were sufficient to kill a duck and a hare, but the meat from these was insufficient. They needed to be able to take musk oxen in order to survive. So they made a lasso from their sledge thongs and lances from whalebone and eventually were able to hunt enough musk oxen to live. But then they had to use these primitive weapons to defend their cave against polar bears who were attracted to the scent of the meat. On November 3, the sun went down, and they were besieged in the dark for a hundred days.

On February 18, the sun came up, and with an outside temperature reading of –40°C, the three left their cave and headed down Jones Sound for Baffin Bay, hauling their musk oxen remains behind them on a sledge. By March 25, their food was gone, but they reached Cape Faraday, only to be caught by a polar bear in the open. Cook had hidden away one last bullet. Knowing who was the best shot in the group, he gave the round to Ahwelahtea, who took down the bear. They then found some seal meat that had been cached at that point by Etukishook's father, and they lived on it until they reached Anoatok.

It is doubtful that Cook actually reached the North Pole. But whether he did or did not, his incredible fourteen-month trans-Arctic trek and feat of Stone Age winter survival on Devon Island stands as one of the most epic adventures in the annals of human exploration.

THE FOUNDING OF POLAR SHELF

With the onset of the twentieth century came the airplane, and the heroic age of polar exploration passed into history. Those who went north now went not for eternal glory but for allegedly more practical purposes.

In the 1950s, a time halfway between the era of Cook and Amundsen and our own, the world was split into two hostile blocks. On the one side was the United States and its allies, on the other was the Soviet Union and its satellite states. In between the two, straddling the most direct line of travel for nuclear-armed bombers and ballistic missiles, was the Canadian high Arctic. The region thus became a place of intense strategic interest.

The United States and Canada responded by setting up the Distant Early Warning system, or DEW line, a chain of radar stations across the entire high Arctic. They also set up airfields in many places, including Frobisher Bay (now Iqaluit, the Nunavut Territorial capital), and Resolute Bay, on Cornwallis Island. Much of the high Arctic was uninhabited at this time, even by Inuit. So to reinforce its sovereignty over the region, the Canadian government forced the relocation of seventeen Inuit families from northern Quebec to Resolute Bay and Grise Fiord, creating the communities that exist in those locations today.[18] Scientific knowledge of the Arctic was also of potentially important military value, and so departments of polar and Arctic studies were set up at many North American universities, and a Canadian government agency was created, the Polar Continental Shelf Project (commonly known as Polar Shelf) to provide logistic support for scientists working in the field. A major base for Polar Shelf was set up adjacent to the military airfield at Resolute Bay, about 180 kilometers from the crater on Devon Island.

Thus was born the great age of postwar Arctic research.

So it was that flying out of the Polar Shelf base in Resolute Bay in the 1950s, scientists first saw Devon's crater from the air. H. Griener, a geologist working for the Geological Survey of Canada (GSC), scouted it and identified its circular shape. He did not realize its true origin, however, and interpreted it as a salt dome. Thus mistaken, he dubbed it Haughton Dome, after the nineteenth-century British naturalist Samuel Haughton. The name was one to conjure with. Haughton had been a friend of M'Clintock and Jane Franklin. When M'Clintock retuned from the expedition that discovered Sir John Franklin's fate, he also brought with him a case of rock samples. Haughton used those samples to write the first geological description of the high Arctic, thereby becoming the founding father of Arctic geology.

It was not until the 1970s, however, that anyone guessed what Haughton Dome really was. The first to put forward the correct hypothesis was Canadian impact geology pioneer M. R. Dence in 1972, who based his supposition on the object's unique shape and form (or morphology, to use the geologists favored five-dollar word). Dence's theory that Haughton Dome was a crater resulting from a large prehistoric meteor strike on the Earth was initially controversial, but was strikingly confirmed in 1975 by GSC geologists P. Blyth Robertson, G. D. Mason, and Richard A. F. Grieve, who discovered shatter cones and other shock-fractured rocks whose origin could only be explained by an impact event.[20, 21] The formation was therefore renamed Haughton Crater.

In the late 1970s, knowledge of Haughton Crater was advanced much further by T. Frisch and R. Thorsteinsson, both of the GSC, who carried out the first in-depth geological studies of the formation.[22] Interest in impact events was greatly heightened after 1980, however, when the team of Nobel Prize–winning physicist Luis Alvarez and his paleontologist son Walter Alvarez put forward strong evidence supporting the theory that the huge mass extinction that terminated the dominance of dinosaurs on Earth 65 million years ago had been brought about by the death-dealing impact of a large asteroid.[23] (The outlines of the crater left by that impact were subsequently discovered by Canadian geologist Alan Hildebrand in the Yucatán in 1991.)

With the importance of impacting asteroids to Earth's history thus made so dramatically clear, it was decided to set up a special international program to advance the scientific understanding of Devon's crater and its environs. Dubbed the Haughton Impact Structure Study (HISS) Project, this program brought together teams of paleontologists, geologists, and geophysicists from the United States, Germany, and Canada to research the physical, geological, and biological history of the crater before, during, and since the impact event. The findings of the HISS project were eventually published in 1988, and collectively represent one of the most comprehensive interdisciplinary studies of an impact crater ever performed.[24, 25]

Thus, by the early 1990s, Haughton Crater was well known to science.

MARS RESEARCH BEGINS ON DEVON

In 1996, Pascal Lee was a young man in a hurry. Born in Hong Kong of Chinese and French parents, Lee had been educated in France, and then received a doctorate in planetary science from Cornell. He wanted to be involved in the U.S. space program, but as a French national, getting hired by NASA or other American organizations was difficult. Lee had some soft-money postdoctoral support from the National Research Council, but if he was to stay in the United States, he really needed a job. To get a job, he needed a project.

Lee was well aware of the studies of Mars analog environments that were being done by Chris McKay and others in Antarctica and elsewhere. Much of Mars's geology has been shaped by meteoric impacts, however, and there were no proven impact structures in Antarctica. But there was one in the high Arctic—the well-documented Haughton Crater on Devon Island. Lee approached McKay, by then a central figure in Earth analog Mars studies at NASA Ames Research Center, and made a case for the need to investigate a polar impact crater as part of the program of Mars analog research. McKay thought Lee's points were well taken, and arranged a $20,000 NRC grant to allow Lee to travel to the Arctic to see what he could find.

So in August 1997, the expedition headed north for Devon Island. Given the modest funding available, the group was kept to a party of five, consisting of Lee, NASA Ames geologists Jim Rice and Aaron Zent, John Schutt, a geologist and experienced field guide for the U.S. Antarctic Search for Meteorites program, and a Greenland husky named Bruno, who was hired for the season from an Inuit family in Resolute Bay. The small size of the team was not an impediment to getting work done, however. In fact, probably in significant part because of its manageable dimensions, this expedition proved highly productive.

The team explored Devon both on the ground and from the air, identifying a number of important parallels between Devon and Mars.[26] In addition to the obvious fact that Haughton was an impact crater, and therefore a geological formation of a type that exists in large numbers on Mars, there was the derivative fact that many of the rocks in and around the crater would have had their morphology and even their composition altered accordingly by shock-induced fracturing, outgassing, melting and recrystallization, or vaporization followed by condensation. These rocks had then been processed over time in a cold environment whose average yearly temperature of –20°C (–4°F) was not too different from the –55°C (–67°F) prevailing on Mars. Thus the geology of the area was potentially uniquely Mars-like. The morphology of the crater was of interest, as it could be dated fairly precisely to 23 million years ago. Partially eroded yet largely intact, it is in a comparable state of preservation to Martian craters dating back perhaps 3 billion years. From this one might draw the inference that erosion rates on Mars have been proceeding at a rate on the order of 100 times slower than those prevailing in Earth's Arctic.

There were polygons and other ground-ice features visible on Devon that showed clear resemblances to phenomena imaged by the Mars Global Surveyor (MGS) spacecraft in orbit around the Red Planet. There were also gullies, water-channel networks, and canyon systems that also bore striking similarities to those seen by MGS on Mars. These results were interesting. The polygons on Devon are caused by ground ice which the team determined sometimes constitutes up to 80 percent of the highly fractured ground. Could Mars's soil have equivalent water content? (The team scored a bulls-eye with this guess. Neutron spectroscopy measurements subsequently taken by the Mars Odyssey orbiter in 2002 indicate that in parts of its polar regions, Mars's soil water content exceeds 60 percent.) According to the team, Devon's canyons were created by glaciers. Could the giant canyons of Mars have been carved in the same way? (That would be a surprise. Most other Mars geologists think the Martian canyons were formed by sapping.) According to the team's findings. Devon's weird Mars-like water-channel networks were caused by water melting under ice sheets in a frigid ambient climate. Could some of Mars's similar channel networks have been created in the same way? Clearly they could. Could they *all* have been? If so, then one of the primary pieces of evidence for a warm-wet Mars in the distant past might be void. Lee went a bit out on a limb arguing for this radical interpretation, because the early warm, wet Mars theory is also supported by unignorable evidence of a past northern ocean on the Red Planet, but still, his data were intriguing.

In short, the team had found a new and unique Mars analog environment that clearly merited further investigation.

On their last day on the island, the team stowed all their gear in compact form and waited for a Polar Shelf Twin Otter to fly in from Resolute Bay to pick them up. It's good to have all your equipment packed tightly if you wish to be able to pack an entire summer expedition's equipment into a small airplane. Unfortunately, in doing so on this occasion the team managed to place their shotguns inside of packed boxes in such a way that they became completely inaccessible. This, it must be said, was rather dumb, as was irrefutably demonstrated a short while later when an adolescent polar bear chose this moment to make his appearance.

Polar bears are extremely dangerous. Their natural habitat is the sea ice, where they live by hunting seals. In the summer the sea ice melts, and they are marooned ashore, where they must fast for months until the ice returns. An adult polar bear can weigh over a tonne and run at speeds exceeding 30 miles per hour. They have eight-inch claws and are intelligent enough to stalk prey, including human prey, from downwind. Michael Crichton's velociraptors have nothing on polar bears. Polar bears are bigger than velociraptors, stronger, smarter, faster, more resistant to cold and bullets, and amphibious to boot. Furthermore, they are real. Every year in the Arctic, several people are killed by polar bears. They are the top land predator on the planet. Even an adolescent polar bear is more formidable than a typical full-grown grizzly. In packing away their guns, Lee's party had made themselves helpless.

Fortunately, Bruno knew what to do. He *charged* the bear. Now, no dog, even a big strong husky like Bruno, can actually defeat a polar bear in single combat. But Bruno's onslaught was so fierce and malignant in its aspect that the bear was bluffed. He backed away. Bruno followed, and barking and snarling and nipping at the young giant's heels, hounded the bear up a wall of breccia and chased it right out of the crater.

For this reason that particular wall of the crater is now known as the Bruno Escarpment.

The report of Lee's expedition created significant interest in the Mars research community. French planetary scientist Nathalie Cabrol and British Antarctic Survey microbiologist Charles Cockell were intrigued. Mars was cold, but the energy released by a large impact hitting the planet could create locally warm conditions for thousands of years. This could be more than long enough for a bacterial ecosystem to develop and flourish, perhaps to send out spores that would litter the planet in wait for another life-promoting impact elsewhere. Could Devon Island have been the host to such impact-induced hydrothermal ecosystems? The ultraviolet on Devon was higher than in the Antarctic dry valleys. Could Devon endoliths have therefore developed protective adaptations even more Mars-capable than those in the Antarctic? In the Antarctic and the Negev, endoliths had only been found in soft rocks like sandstones. But on Devon, there were hard crystalline rocks that had been shattered at the microscopic level by the impact. Could the endoliths there have learned to exploit such places?

NASA's technologists were interested too. If Devon was so Mars-like, perhaps it could be used as a testing ground for various technologies of potential use for Mars explorers, such as advanced communication equipment, exploration robots, and ground-penetrating radar. The Antarctic was already being used for such purposes, but getting to Devon was considerably cheaper.

Support for Lee's project was increased to $80,000. A new expedition of some thirteen people, including several biologists and technologists, was planned for the summer of 1998.

Bruno was rehired.

5. TOWARD FLASHLINE STATION

In April 1998, as the Mars Society was moving toward its founding, a meeting was held of the core group of its nascent Steering Committee at NASA Ames Research Center. Present were Chris McKay, Carol Stoker, Larry Lemke, Geoff Briggs, and I. Our purpose was to try to define a suitable initial project for the Mars Society to undertake with its own resources.

I advocated a Mars balloon mission. I had worked on designs for Mars balloon missions before,[27] and I believed that if we cut the gondola payload down to just a camera and adopted a minimalist design philosophy in other areas, we could reduce the trans-Mars injection mass of the entire balloon system and its entry capsule to less than 20 kilograms. Such a small unit could therefore be flown as a hitchhiker payload on one of the many NASA or ESA Mars probes planned for the next decade. No one had ever flown a balloon or any other type of aircraft on Mars, and seeing the planet from a bird's-eye view would have real scientific value. More than that, however, it would drive home to the public the profound reality that there was a vast new world waiting for us to explore.

Larry and Carol favored a project to build a small solar sail spacecraft and fly it on a cycling trajectory between Earth and Mars. The concept of solar sails,[28, 29] in which a spacecraft is propelled without fuel by deploying a large thin foil to catch the pressure exerted by sunlight, has been known to the astronautical literature since the 1890s. Yet due to lack of sustained vision within the world's space agencies, no such device has ever been flown. Because they require no propellant, solar sails are indefinitely reusable, and could someday represent an economical means for transporting cargos between Earth and Mars. More futuristic ultrathin solar sails might even eventually become an enabling technology for interstellar travel. The Mars Society could make a major contribution to the progress of human expansion in space by being the first to demonstrate such a technology.

Chris's ideas were less definite, but whatever we flew, he wanted it to focus on the question of life. Life was the central issue of concern in thinking about Mars, and we needed to address it. Our project therefore should be some sort of hitchhiker delivered to the Martian surface, carrying a payload that either served as some kind of life-detection experiment or, alternatively, demonstrated the transportation of life to Mars from the Earth. For example, we could deliver some seeds to Mars, and grow a flower inside of a contained chamber. This would be a symbolic first step on the path toward bringing the Red Planet to life.

The problem with all these ideas was that they required launching payloads into space. Space launch would either require paying very large amounts of money, which we did not have, or negotiating a deal with someone else who would provide us with a hitchhiker opportunity at low or no cost. Such deals are very difficult to negotiate, and furthermore, should the primary payload on such a mission experience unexpected mass growth, the hitchhiker ends up getting jettisoned. Thus years of work can go for naught. This problem was even worse for the mission plans proposed by Chris and me, since we not only had to hitch a ride to *space,* we needed a ride to *Mars*. This greatly reduced the number of useful spacecraft potentially available, and furthermore cut the frequency of opportunity to once every two years. Starting a project in 1998, we were already too late to go along on the Mars probes scheduled for 1999 or 2001. Our earliest chance would be 2003, and if we were not successful in moving our program very quickly, we would have no chance to fly before 2005. This would be a very long time for a membership society to have to wait before seeing results from its activity.

The concept of the solar sail mission did not suffer as much in these respects, but it still required getting a launch to a high Earth orbit (since solar sails experience too much aerodynamic drag in low Earth orbit to propel themselves into interplanetary space). Moreover, other groups and technical societies were already pursuing private demonstrations of solar sails, and so our project would not be original, and might even be seen by some as competitive.

The meeting thus ended without satisfactory resolution. But as I exited the building to go back to the airport, a Corvette screeched to a halt in front of me. The sports car opened, and out popped a young man of Eurasian appearance. He introduced himself as Pascal Lee. He asked if it were true that we had just had a meeting to determine the Mars Society's first project. I said, yes, and told him about some of the concepts in play.

Lee said he had a better idea. Instead of initiating a project with a problematical requirement for space launch, why not build a human Mars exploration analog research station on Earth? He had found the ideal place for such a station: Haughton Crater on Devon Island. The purpose of the station would be to continue the geologic exploration of Devon, but to do it in the same style and under many of the same constraints as would be involved in conducting such activities on Mars. By doing so, researchers would be forced to confront some of the real problems of human Mars exploration and begin the process of developing appropriate field tactics.

I liked the idea instantly. A Mars Arctic Research Station could be done more quickly and cheaply than any of the projects we had been considering so far. Moreover, it was specifically geared toward the issue of human Mars exploration, which took it completely outside the intellectual framework of the established NASA robotic Mars exploration program. The concept of a human Mars exploration practice station was not original. Ben Clark had developed a design of an Antarctic manned Mars research station as part of the Martin Marietta Scenario Development Team activity in 1989–90, and papers calling for the establishment of such a station had been published by numerous authors going back decades. But no one had ever built one.

Furthermore, there were no plans to do so now. NASA was prohibited from spending money on such a project by congressional opponents of human Mars exploration, because they saw it as the camel's nose in the tent that could lead to an actual humans-to-Mars program.

Yet such research is vitally necessary. For example, it is one thing to walk around a factory test area in a new spacesuit prototype and show that a wearer can pick up a wrench; it is entirely another to subject that same suit to two months of real fieldwork. Water use is a key variable in defining Mars mission logistics requirements, yet is unknown, and will remain unknown until assessed in the context of a program of active field exploration. Psychological studies of human factors issues, including isolation and habitat architecture, are nearly useless unless the crew being studied is attempting to do real work. The impacts on crew operations of candidate subsystem maintenance requirements, or various proposed procedures, can also only be measured if the crew is really operating.

Moreover, there is an operations design problem of considerable complexity to be solved if human Mars exploration is to be done effectively. A human Mars mission will involve diverse players with different capabilities, strengths, and weaknesses. These include the crew of the Mars habitat, pedestrian astronauts outside, astronauts on unpressurized but highly nimble light vehicles operating at moderate distances from the habitat, astronauts operating at great distances from the habitat using clumsy but long-endurance vehicles such as pressurized rovers, mission control on Earth, the terrestrial scientific community at large, and robots. Taking these different assets and making them work in symphony to achieve the maximum possible exploration effect will require developing an art of combined operations for Mars missions. The Mars Arctic research station program would begin the critical task of developing this art.

The Mars Arctic station plan had other advantages. By focusing on developing techniques for manned exploration, it would put the vision of human expansion into space foursquare before the public in a way that no robotic program like the balloon mission or the solar sail spacecraft possibly could. The fact that we would do so in the Arctic had a resonance too. I was well aware of the history of Arctic exploration, and believed that the lessons learned from the experiences of Franklin, Amundsen, and the rest have relevance for those who would plan future Mars missions. It was a study of Amundsen's techniques, for example, that provided part of the basis for the travel-light-and-live-off-the-land design philosophy underlying the Mars Direct plan. In implementing a program of human Mars exploration training in the Canadian Arctic, we would be carrying forward a tradition well worth honoring.

The politicians could hamstring NASA, but they could do nothing to stop a private organization from creating such a station. The Mars Society could do what NASA could not. We could make the vision of a Mars Arctic Research Station real.

The idea was so compelling that within weeks every member of the Steering Committee was in agreement that the Mars Arctic Research Station (MARS) should be the Society's first project. At the Steering Committee meeting at the Founding Convention in August, the MARS project was formally adopted. When the decision was announced to the membership at the Saturday night banquet town meeting, it was greeted with acclamation.

Virtually immediately, volunteers stepped forward to help make it happen.

THE ARCHITECT APPEARS

The first and possibly most important of the Mars Society volunteers to get involved in the period immediately following the Founding Convention announcement was Kurt Micheels. A short, stocky man, Kurt was an architect based in California. He also had construction experience, and as a Lieutenant Commander in the Naval Reserve, some background in operations logistics. He cornered me on Sunday, the final day of the conference. We had a long talk out on the patio in front of the student union building where the convention was being held. He told me about his credentials and said he wanted to help. As an example of what he could do, he took out his pad and started sketching designs for the station.

The collaboration continued from there. Kurt would make sketches and fax them to me, after which I would return them with comments or call him up on the phone to discuss them. Kurt explored a number of alleys, and by October he had come up with a design that I considered good enough to put out publicly as illustrating our basic concept. It was a domed cylinder twenty-seven feet in diameter and twenty-eight feet tall, fitted out with three decks. The upper deck contained the crew staterooms. The middle deck contained the communication areas, the laboratory, and the wardroom. The lower deck contained the workshop, EVA preparation areas, and the airlocks. There were toilets on each deck.

The twenty-seven-foot diameter Kurt chose is a magic number. It is the maximum diameter for cylinder fabrication of the Lockheed Martin facilities in Michaud, Louisiana, that are used to fabricate the Space Shuttle external tanks and which were previously used to make the lower stage of the Saturn V moon rockets. For that reason, the hab designs developed for the Mars Direct plan and the NASA JSC Design Reference Mission both also used twenty-seven-foot-diameter cylinders. In Mars Direct, however, the cylinder was only sixteen feet tall and fitted out with just two decks. This was in keeping with the Mars Direct plan's spirit of minimalism and the fact that it only employed a crew of four. We had decided that the

MARS hab would support a crew of six. Going from two to three decks therefore seemed reasonable.

In parallel with and supporting Micheels's architectural design effort, an Arctic Task Force was formed. This was a group of Mars Society volunteers including Micheels, an Air Force Captain named Tom Hill, and many others. The Arctic Task Force started working through the issues of hab power supply, water system, waste disposal, communications, and potential options for transporting the station to Devon Island once it was ready for deployment.

This last item was a major issue. Because of the anticipated difficulty of engaging in construction on Devon Island, some proposed building the habitat in Resolute Bay and then transporting it in one piece from Resolute Bay to Devon using a giant Russian lifting helicopter. It was pointed out that bringing the hab in that way would have a striking resemblance to an actual Mars mission. As one person put it, "What a picture that would be!" But an investigation of the cost of such a procedure ultimately forced the Arctic Task Force to reject it in favor of on-site construction.

FINDING THE CASH

The key requirement that needed to be addressed if the MARS project was to become a reality was money. My rough guess at that time was that the station would cost somewhere between half and million and a million dollars. The Mars Society had $30,000 in the bank. Fundraising was the central task at hand.

I spent a lot of time pursuing dead ends. In February 1999, however, I got a promising lead from an unusual source: the barbershop. I was sitting there, waiting for my turn, and decided to kill time by reading a copy of *Time* magazine. In the much-abused issue, I found an article under the title "The New Philanthropy." It was all about the new rich, the internet dot-com billionaires. These whiz kids did not donate to the Junior League or the Opera. They grew up watching *Star Trek,* and were into more futuristic causes. One of them, Infoseek Chairman Steve Kirsch, for example, had just made a major donation to fund a search for Earth-impacting asteroids.

It seemed to me that Mr. Kirsch could be well worth talking to. The Arctic Task Force was having a meeting in the Bay Area in a few weeks, and I was planning to attend. Kirsch lived in Palo Alto. I contacted him and arranged a meeting to coincide with the trip.

I went to Kirsch's house and started to brief him. He did not take me seriously. He took a copy of the Mars Society brochure and looked at the list of Steering Committee members listed there. "Chris McKay? Mike Griffin? Who's Chris McKay? Who's Mike Griffin? I've never heard of Chris McKay or Mike Griffin."

I explained to him that these people were well known in the space-exploration field, but he would have none of it.

"That may be, but I've never heard of them. If you want people to donate, you need a board with names that people have heard of. You need someone like Buzz Aldrin. If you brought Buzz Aldrin to my door, I'd give you a million dollars."

"Done!" I said with alacrity, sticking out my hand to shake on the deal.

Kirsch took two steps backward. "No, no," he said, suddenly cautious. "I spoke too soon. One hundred thousand dollars. If you can bring Buzz Aldrin to my door, I will give you one hundred thousand dollars."

"Done," I said again, stretching my fingers to indicate it was time to shake.

Kirsch took my hand and we shook. "Remember," he said, "you really have to bring him right here or there will be no money."

"Understood," I said.

I left the house happy. Kirsch probably doubted I could fulfill my end of the bargain, but in point of fact, I had known Buzz Aldrin for years. While many of the Apollo astronauts had simply used their fame as a platform to advance themselves in other careers, Buzz recognized that what he had done on the Moon would be reduced from a historic milestone to a meaningless stunt if there were no follow-through. He had therefore dedicated much of his time and effort over the years to promoting human space exploration. The two of us had worked together as officers of the National Space Society. I felt sure he would help now.

I called up Ron Jones, the man who was arranging Buzz's affairs, and asked him to relay my request. He did, and a day later the answer came back positive. Buzz would lend a hand. That was great. The problem now was to actually arrange a meeting. Kirsch's schedule was tight. Buzz's was insane.

For the month of March, Buzz's schedule was something like this: On the tenth, Bangkok. The eleventh, Irkutsk. The twelfth, Reykjavik. The thirteenth and fourteenth, Buenos Aires, then Nashville on the sixteenth, Fiji on the seventeenth through the nineteenth, Mexico City on the twentieth, London on the twenty-first, and so on. I'm not kidding. The only free day he would have in California was on the thirty-first. After that, he would not be available again until October.

I called up Kirsch and asked if he was available on the thirty-first. He said he would be. That was great. But then two weeks later, Kirsch e-mailed me and said something had come up. Could we postpone for a week?

I would have been happy to postpone for a week, except that was not possible. The thirty-first was the only opening for six months. I certainly did not want to postpone until October—the whole deal might flake out by then. So I told Kirsch it had to be the thirty-first. He probably thought I was being very rude; surely a week's delay was not too much to ask, especially since it was his $100,000 after all. But I had no choice. Miraculously, he agreed. The meeting went off as planned. Buzz and I had a nice Italian dinner with Steve and Michelle Kirsch, and afterward he told me he would have his charitable foundation mail us a check.

This was a good start, but I needed more. As soon as I had the commitment from Kirsch, I called up Rick Tumlinson. Tumlinson was the president of a small space-advocacy organization called the Space Frontier Foundation (SFF). The SFF's claim to fame is that it believes only in privately funded space activity. Some of its members are anti-NASA, because it is a government agency, and anti-Mars as well, since they believe that Moon bases and solar power satellites are more viable venues for commercial activity. Tumlinson was not so extreme. He was willing to support privately funded Mars-related activity. Moreover, he had at his disposal the means to do so. An eccentric libertarian telecommunications billionaire named Walt Anderson had created a trust called the Foundation for the International Non-governmental Development of Space (FINDS), endowed it with $30 million in stock, and entrusted Tumlinson to run it for him. Tumlinson had attended the Mars Society Founding Convention, and upon hearing about our MARS project had said that "once we really got it going" FINDS would be willing to add its support to the project with a donation "in the fifty to one-hundred thousand dollar range."

Nothing had come of this pledge so far, but now we actually had a firmly committed donation large enough to put the MARS project in motion. I told Tumlinson we had $100,000 on the way, and that it would have the most impact if we could announce both donations together. He said, "All right, I'll send you $50,000." I replied, "Do you really want to be the lesser of the two donations?" Tumlinson thought about it for a few seconds, and then said, "Okay, we'll send $100,000 too."

Both Kirsch and Tumlinson honored their pledges. By mid-April 1999, the MARS project had $200,000 in the bank. We were finally in a position to really get started.

THE SCOUTING EXPEDITION

The next step was to find a site for the hab. Pascal Lee's thirteen-person 1998 expedition to Devon had been successful, and in consequence his funding had increased to about $200,000, allowing him to plan a field season for 1999 involving some thirty people, with roughly twenty on the island at any one time. The program, now grandly dubbed the NASA Haughton Mars Project, or HMP, was actually a kind of potluck affair with different NASA or university scientists dropping in and conducting their own research out of Pascal's base camp. This time, however, the Mars Society, in the persons of Kurt Micheels and yours truly, were to be among the participants. (Actually most of the people at the HMP camp, from Lee on down, were Mars Society members. But Kurt and I were there on Mars Society business.) We went to Devon to learn enough about the place to plan operations and to determine an optimum site for the station.

Proceeding to Devon in late July, when the weather could be expected to be optimal, we obtained our first introduction to the realities of air travel in the Arctic. Our plane from Yellowknife to Resolute was delayed for unexplained reasons for thirty-six hours. Instead of a comfortable Saturday afternoon flight, we were rescheduled on one leaving midnight Sunday, flying due north right into the sun. This landed us, at around 3 A.M., in desolate Resolute Bay (72 N), where we were brought to the Polar Shelf dormitory and told to be ready for the Twin Otter that would take us over to Devon around 10 A.M. Fortunately, our biological clocks made us rise early, so we were awake and able to catch the Twin Otter when it actually left at 8 A.M.

Twin Otters are the workhorses of high Arctic travel. They have two propellers, which are mounted high, allowing the planes to operate off of very short and rough airstrips. They have a crew of two, and seat seven people, or alternatively, can carry a cargo of up to 1000 kilograms. In this case, the plane carried five. In addition to Kurt and me, there was also NASA's Carol Stoker, the Canadian Space Agency flight surgeon Gary Gray, and George Dunfield, a geotechnical expert for the HMP.

As we flew from Resolute to Devon, we could see the massive packs of sea ice, which was beginning to crack into channels of open water, as the late summer did its transient melting. Then we were over Devon, and looking down, could readily make out the networks of eroded channels, which Pascal had cited for their uncanny resemblance to features imaged on Mars. Finally, we flew over and into the crater, whose immense round shape and circular gray-powder walls left little doubt, to me at least, of extraterrestrial origin.

The plane landed in the crater right next to the HMP camp, which was a group of three large and about twenty small tents positioned by the banks of the Haughton River. The whole camp turned out to greet us and help unload the plane. Then, after we pitched our tents, everyone was assembled for a briefing by Pascal. As this briefing was my first, I found it quite interesting, a feeling that was to diminish somewhat as it was repeated many times with minor variations over the following days.

After completing lengthy introductions and explanations of camp policies, Pascal went over everyone's planned activities for the day, identifying who would have each All Terrain Vehicle (ATV), where it would go, and for how long. All of us in the new group were chosen to head out with Pascal and several others to examine the several candidate sites that HMP people had identified as possibilities for the Mars Arctic Research Station. Before we left, however, we were given another briefing on the handling of the anti–polar bear weapons (12-gauge pump-action shotguns loaded with heavy steel slugs instead of birdshot) and the ATVs by John Schutt. John looked the part of an Arctic guide, with his mountain-man beard, but conversation soon revealed that in addition to being a crack Arctic hand, he is also a very knowledgeable geologist. After some training with gun and vehicle, we had lunch and then set out.

The ATV excursion was wonderful. The weather was cloudy and cool, perhaps in the high thirties, and the landscape was alien; lifeless but strangely beautiful. As we set out through it, our spirits were the highest imaginable.

My diary for the day, July 26, 2000, reads:

> We traveled up the Bruno escarpment. . . . There was a possible station site at the top, but for many reasons it seemed unappealing to me. Then Pascal took us to another unsatisfactory site. After a trip of about 45 minutes, we arrived at a third site, a ridge overlooking the north edge of the crater. The ridge is covered with small sharp boulders and looks a lot like one of the Viking landing sites. There is a small valley north of the ridge, through which a small stream flows, after which there is a very flat plateau which appears to be serviceable as a landing strip for twin-otter aircraft. Beyond the potential landing strip is a huge plain, which Pascal has named the Von Braun Planitia. The view from the ridge where we stood was spectacular, looking out to the south to

> the crater and to the north to the planitia. Locally the geology is very Mars-like, and the nearby sources of water and adjacent landing strip make it potentially very attractive for the MARS base.

There were some issues with potential difficulty of construction on the ridge site, so we kept searching for a better one, via ATV or on foot, for the next four days. As things turned out, however, we could not find a superior alternative, and the site was selected for the MARS. I named the ridge Haynes Ridge, after Professor Robert Haynes of York University, a founding member of the Mars Society, recently deceased, who had been a seminal thinker on issues concerning the terraforming of Mars. The river that separated Haynes Ridge from the plateau airstrip was named the Lowell Canal, after the nineteenth-century astronomer who thought he saw artificial waterways on Mars. Both names stuck, and the features in question have been referred to by these terms ever since.

We returned to the base through a system of canyons. In the course of this trip, out and back, we took the ATVs over terrain that I would not have believed wheeled vehicles capable of, covering steep slopes or sharp rocky boulders with great facility. Arriving back at the camp, Carol Stoker and I agreed: ATVs (or their functional equivalent) are a must for human Mars missions. The ease of travel over very Mars-like terrain, the close visual interaction with the environment, and the informality with which side trips can be undertaken to follow up on visual clues to interesting discoveries, will make such systems a tremendous multiplier of the scientific return of any human Mars expedition. Also, without question, the pilots among the astronaut corps will find them a blast.

Everyone at camp had dinner together, prepared for all by two who remained behind during the afternoon sortie. Afterward people hung out and talked till quite late. On Devon in the summer, the sun does not go down, and it takes a conscious decision to retire to sleep. But when I finally got to my tent about one A.M., I slept like a rock.

Over the next several days, the pattern of the first was repeated, with the main variation being evening entertainment. We would have breakfast, followed by Pascal's briefing, then there would be a scouting ATV traverse or a hike that would occupy most of the day, followed by dinner together at camp, and then another briefing by Pascal.

After Pascal's after-dinner briefing, he would usually arrange a program of after-dinner entertainment, consisting of a series of lectures followed by a movie. The lectures had considerable variety, as the camp included about twenty people who were leading experts in one field or another. The movie, however, was always the same, as the only DVD movie anyone had brought to the island was the farcical spy flick *Austin Powers.* Everyone at camp had watched this thing again and again, and many had memorized it. Austin's expression "Shagadelic Baby!" was a phrase on everyone's lips, creating a pleasantly demented conversational atmosphere.

There were other amusing sidelights to the scouting expedition. Here is the report for my diary of July 28:

WEDNESDAY, JULY 28, 1999

The ATVs were all taken by others, including two by Charles Cockell and a grad student named Oz [Gordon Ozinski]. Charles is an exobiologist from the British Antarctic Survey. On one of his recent sorties he detected evidence of fish in a small lake that had been isolated geologically from any larger body of water for at least 10,000 years. Charles announced that he is going to try to catch one, in order to compare it to similar species to study genetic drift. His equipment was primitive: he had some fishing line and a hook, but no rod or reel, lures, bait (other than old sausage), or knowledge of the appropriate technique for the fish in question. No one thought he would catch anything. But the season is nearing its end, and for scientists like Charles who have been here more than a month searching for (and finding) endolithic organisms, it was time for a lark.

Without ATVs we couldn't go far, but Kurt, Carol, I, and a NASA scientist named Bill Clancey went for an extended hike through the crater, in search of potentially better spots than our base camp should the ridge prove unsuitable. We didn't find any place that is significantly better, but we did come across some spectacular gypsum crystals, including some massive outcrops several feet across.

On our way back in the afternoon we met Charles and Oz returning from the lake. They had caught a huge Arctic char, about two feet long and probably weighing close to 20 pounds. At camp that night Charles received repeated hints that only the fish's head is needed for science, and the rest of it might better be dealt with by us. He resisted all such entreaties, however, and the fish was carefully packed for shipment to Resolute.

Upon arriving back at camp, I decided to wash my face in the Haughton River, which is made of Arctic meltwater. This proved to be a very interesting experience, and quite refreshing. I recommended it to others, but got few takers.

After dinner the crew presented me with a miniature model of a Mars Direct habitat lander, complete with miniature tuna-can habs made of—what else—Chicken of the Sea tuna cans! They got me to pose with the thing for the camera. It was a fun moment.

We had more fun on the trip, including an incredible 1 A.M. hike up the Bruno Escarpment to the crater rim. But on Friday evening it was time to leave. The Twin Otter that took us out brought in Lieutenant Colonel Tom Duncavage, a Marine Corps Reserve air mobility officer who had command of a group of C-130 Hercules aircraft. Duncavage was also a mission operations specialist at NASA Johnson Space Center and was interested in seeing what logistic help his Herc unit could lend the Mars project. He and his men had thus come to Devon to paradrop in some supplies for the HMP (thereby also providing training for the Marines), and to check the island out to see if there were any possible landing sites for a C-130. (There turned out to be, but only good enough for emergency purposes. Takeoff was problematical.)

Duncavage and his men would be the ones to do the difficult series of paradrops required to bring the MARS components the following summer. I introduced myself to the officer, and then together with Kurt, Carol, and several others, left the island in his Twin Otter. By the next day, the whole camp had pulled out, and the HMP 1999 season was over.

The bottom-line result of the 1999 scouting expedition was that it was a success. We had found a good site for the Mars Arctic Research Station. The issue now was to build it.

BEGINNING MANUFACTURE

Kurt Micheels and I had plenty of time to talk while we were on Devon. He told me he wanted to go beyond the role of volunteer project architect—he wanted to take on the job of project manager full-time. Now that things were getting serious, we would need such a person if the program was to succeed. He offered to come to Denver and work out of the offices of my company, Pioneer Astronautics, to oversee the design and construction of the station. His commitment to the program was obvious and his salary request was reasonable. I accepted his offer, and in September Kurt moved to Colorado.

Kurt started by working with a Mars Society design engineer named Wayne Cassalls. Together they started turning Kurt's sketches into CAD drawings, suitable for fabrication. It was here that some problems with Pascal started to emerge. Pascal had been appointed Project Scientist, meaning that he had the responsibility for seeing to the successful operation of the station on Devon Island. However, he went beyond this and started to insist on the right to approve or disapprove of all work that anyone else did, including that of Kurt Micheels. While well trained as a planetary scientist, Pascal had no particular expertise in architecture or construction, and his input in these areas was not especially valuable. I had Kurt send Pascal the drawings for his concurrence anyway. Unfortunately, Pascal would generally take weeks before he would look at them. These delays were quite irritating to Kurt, so after a number of such instances, I ruled that henceforth Pascal would have forty-eight hours to comment on any materials sent to him. If he did not comment (he almost never did), concurrence would be assumed and we would move on. This arrangement appeared to annoy Pascal, but there was really very little choice if the program was to make any progress.

The next step was to find a contractor. While we were in Resolute Bay on the way home, Kurt and I had visited with Peter Jess, an Arctic contractor who had given us a preliminary bid of $87,000 to fabricate the primary structure at a factory in Calgary. As this was within our means, it was quite attractive. But then Jess increased his quote to close to double that amount, and then to $215,000. This was too high, and under the circumstances, I was concerned it could get higher. So we decided to put out an open call for proposals for a fabricator.

We received four proposals. One was from an aerospace contractor, who wanted $400,000. Two others were also excessively priced. The fourth was asking $150,000, but the contractor in question was located in New Zealand and wanted all the money in advance. That didn't sound too inviting. So finally realizing what we should have thought of in the first place, I told Kurt to hit the Yellow Pages to see if we could find a fiberglass contractor right in town.

Kurt did a lot of calling and eventually found two promising leads. One company, called Windryder, made fiberglass windmill blades and sailboat hulls. Their capabilities seemed good, but their price was on the high side. The other company, though, had something really special to offer.

The firm's name was Infracomp, for Infrastructures Composites International. It was headed by an inventor named John Kunz who, together with his now-retired father, had developed a unique type of fiberglass honeycomb construction technology. In the Kunz system, two sheets of fiberglass were separated by corrugated fiberglass sections four to eight inches thick, which thereby created a thick honeycomb between the two fiberglass walls. If desired, the honeycomb could be filled with insulation. This created a wall that was waterproof, weatherproof, and corrosion-proof, had good insulating properties (R-20 to R-40, depending on the thickness and the filler), and was very strong and quite light given its strength. A piece of Kunz honeycomb six inches thick was virtually unbreakable, yet weighed no more than ¾-inch-thick pine. Infracomp had great hopes for the material. It could be used to build bridges, or houses. But that was not its true purpose. The Kunzes had invented the stuff to enable a more visionary goal. It was their dream to see this technology used to build floating cities on the open sea.

I liked Kunz, I liked his technology, and I liked the price he quoted: $106,000 to build the entire primary structure, including walls, dome, decks, doors, windows, and legs. The fact that his factory was just across town was a big advantage too—it meant that we could go over anytime and monitor the progress of the work. In January 2000, I signed the fabrication contract with Infracomp. They went to work immediately and had molds produced for all the primary pieces by late February.

The plan was to bring the pieces of the hab into Devon Island by paradrop. Pascal had developed a relationship with Tom Duncavage. Duncavage worked daytime for NASA Johnson Space Center, but as the commander of a reserve air mobility unit for the Marines, he had some influence over what a group of C-130 Hercules aircraft would do to accomplish their required monthly training. Instead of just wasting gas flying around, Duncavage reasoned, why not have the Hercs do something useful for the space program while they were training? In 1999, he had used a Herc to paradrop some supplies into Devon in support of the HMP. In 2000, his planes would perform the paradrop of the components for the MARS station. These would be the northernmost paradrops ever done by the Marine Corps.

This was all great, but the catch was that the station would absolutely have to be ready in time. The Herc flights that brought the station components to the Arctic would also bring north the put-in equipment for the HMP. Departure from Moffett Field, California, was scheduled for no later than July 2. That meant the completely fabricated station had to be on trucks shipping out the factory door for the West Coast no later than June 27. Could Kunz meet that deadline? It was going to be tight.

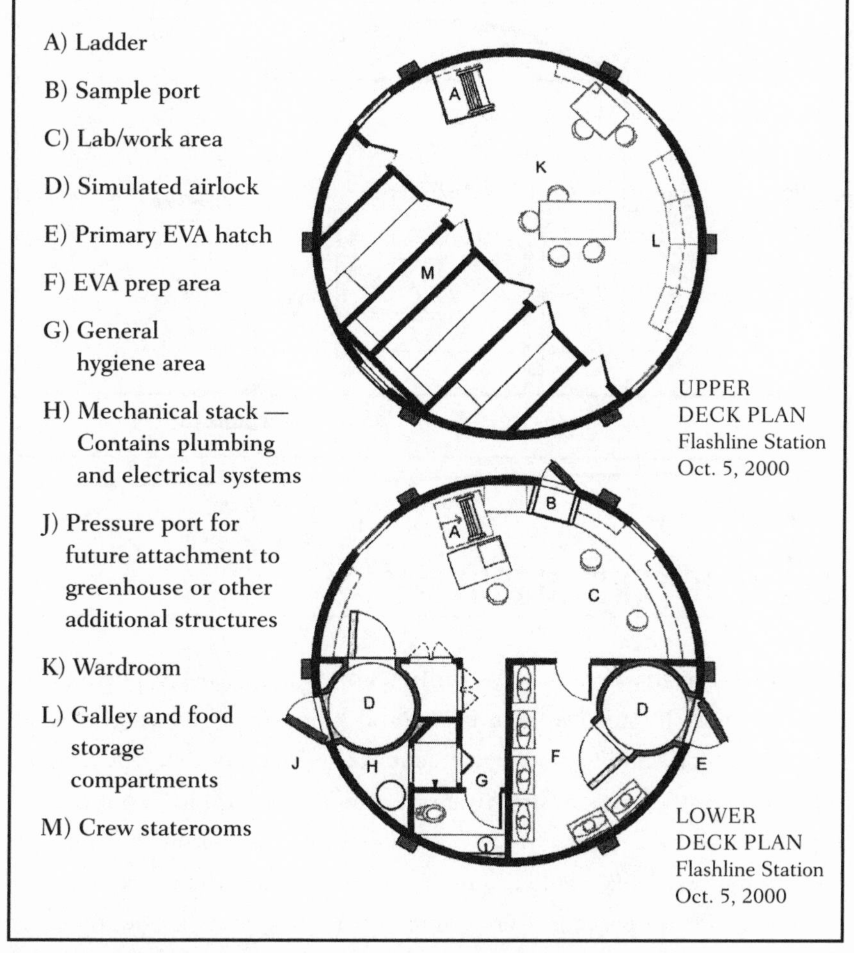

Design of Flashline Station as built. The two-deck design finally selected closely resembles that chosen for Mars Direct and the JSC Design Reference Mission.

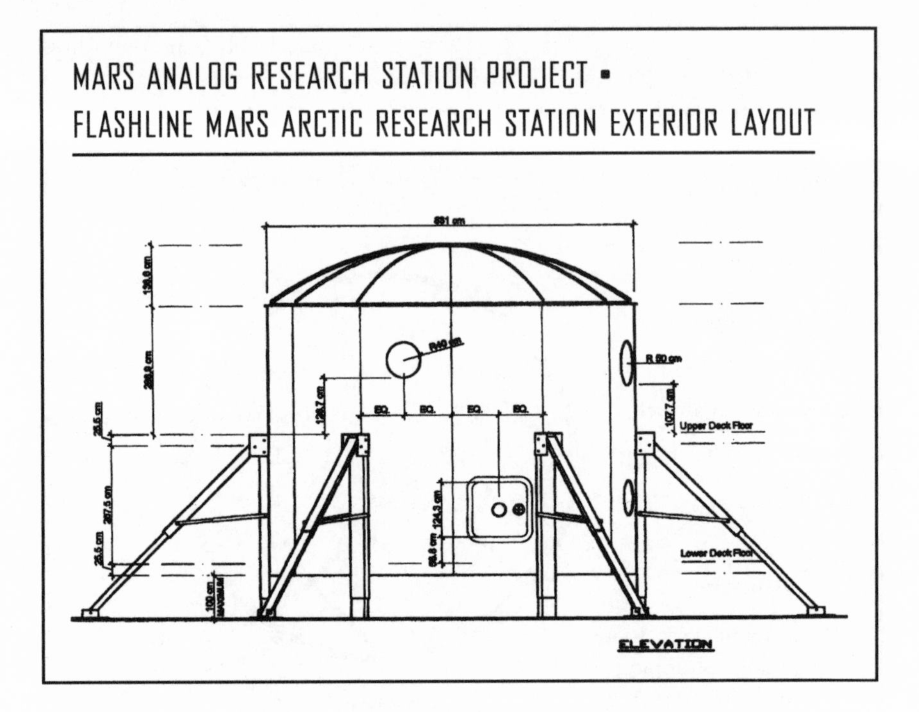

FLASHLINE AND DISCOVERY

With the fabrication under way at Infracomp being closely supervised by Kurt Micheels, I tuned my focus back toward the issue of funding. We had enough cash on hand to pay the Infracomp contract, but that just covered fabrication of the primary structure. We would also have to fit out the interior, transport everything to Devon Island, build the station, and support crew operations. This would require a lot more money than what we had.

I called a lot of people without much luck, until I struck gold in a conversation with Charles Stack. Stack was a wealthy man who had made his money in a company called Bookstacks Unlimited, and was currently the head of a firm called Flashline.com that sold subprograms to software developers. He had been a founding member of the Mars Society, and in fact, through his charity, the Longview Foundation, he and his wife had provided one of the four $5,000 seed donations that enabled the Founding Convention. Now, in early 2000, he offered me a deal. If I were prepared to sell his company the name of the station, he would donate $175,000. One hundred thousand would be paid immediately, the balance a year later. I took his proposal to the Mars Society board, which approved the arrangement. Accordingly, the station became known as the Flashline Mars Arctic Research Station, or variously the FMARS or Flashline Station, for short.

Aside from selling the name, the other major potential source of commercial support for the station was media rights. I knew an independent film producer named Andy Liebman who had filmed me for a documentary entitled *Destination Mars,* which aired on the Discovery Channel to good reviews in December 1996. Andy was also the producer of the Scientific American Frontiers show with Alan Alda, and had featured me in that venue as well. He had won two Emmy Awards for his documentary nature wildlife shows, and was well known and respected by the brass at the Discovery Channel.

I called up Andy and proposed that he do a show on the building and operation of the Flashline Mars Arctic Research Station. Discovery would be the sponsor, paying us a fee for exclusive American TV rights for the first two years, as well as his usual costs for producing a documentary. Andy liked the proposal, and took the idea to Steve Burns, Vice President of Production for the Discovery Channel. A meeting was set up between the three of us at Discovery's corporate headquarters in Bethesda, Maryland, and we had lunch together afterward. Burns and I reached agreement in principle: for $200,000, Discovery would purchase exclusive English-language TV rights to the station's activities for the first two years. Print and radio media rights would remain open.

The essence of the deal thus took only hours to negotiate, but Discovery's lawyers made the process to actually work out a contract take months. They kept on sticking in new clauses that were outside of, or even contradicted, the understanding Burns and I had reached. I was repeatedly forced to recontact Burns to get him to hold Discovery's lawyers to the original agreement. Then the deal was almost shot down when one of the Discovery lawyers thought it would be a good idea to insert a clause explicitly permitting them to avoid filming the name Flashline on the station. This was necessary, he claimed, because when the documentary was actually broadcast, the show might have one of Flashline's competitors as a sponsor. I said this was impossible; neither Flashline nor its competitors ever engaged in broadcast TV advertising, their market was much too specialized for that. But the lawyer would not relent. I felt I could not sign such a contract without Flashline's permission, and they, when asked, were understandably unhappy and would not give it. So I worked out a compromise where we stated exactly where Flashline's name would appear on the hab—right above the front door and to the left—thus providing Discovery with the assurance that it was possible in principle to film the station without the name Flashline always being in frame. I went to Andy and got his assurance that he nevertheless would, in fact, film Flashline's name in many included shots, thereby protecting their interests.

That should have brought the contract to closure, except that Pascal chose this moment to act up. He had taken offense at the Flashline deal, threatening to resign because he had not given his approval to the wording of the press release announcing it, but that episode had only lasted a couple of days. Now he dug his heels in. It was unethical to sell media rates, he said, because it was unfair to all the poorer press outlets who could not afford to pay. As a government agency, NASA could never collaborate with any organization that funded itself through the sale of media rights. Furthermore, it was undignified for a scientific organization to fund itself through media rights sales, and would cast a shadow on the quality of the work done. He absolutely forbade any such arrangement, on pain of complete termination of cooperation between the HMP and the Mars Society.

All this was nonsense. Scientific expeditions and archeological digs are funded all the time by media rights sales, and in point of fact, NASA itself has sometimes sold media exclusives. But Pascal was so adamant about the matter that a meeting had to be held at NASA Ames with management and the public relations department to allow things to proceed.

Eventually, however, Pascal came around. In fact, the HMP made money off the Discovery Channel by charging all sorts of fees to their people for logistic support while they were on the island (totaling more than $150,000 over the two years of the contract).

This cleared the last obstacle to the Discovery Channel deal. The contract was brought to the Mars Society board for a vote, with all approving except for Marc Boucher, the head of the Mars Society Canada, who voted against it. In May of 2000, the deal was signed.

Adequate funding for the Flashline Mars Arctic Research Station project was finally in place.

ENTER THE ROCKY MOUNTAIN MARS SOCIETY

While this was going on, another group of people had become active in a way that was ultimately to prove of critical value to the program. This was the Rocky Mountain Mars Society (RMMS)—our Colorado chapter—led by a core group consisting of Tony Muscatello, Lorraine Bell, Dewey Anderson, Brian Enke, Jason Held, and Robert Pohl, and including many others. These people were all volunteers, who since the Founding Convention had implemented various activities, especially public outreach and political work, in support of the aims of the Society. In the spring of 2000, they decided to start working out communications protocols for the missions that would be carried out at Flashline Station. The way they would do this was by sending a crew on a weekend excursion to do some geology exploration into the Rockies, using Brian Enke's home, which is in the mountains, as a base. They then would communicate via email with another team that was acting as Mission Support in a set of offices we had set aside for this purpose at Pioneer Astronautics. Each side would wait ten minutes before answering any e-mail, to simulate Earth-Mars radio time delay. The primary object was to find out what kind of reports were necessary to actually inform Mission Support adequately of the activities, operational status, and scientific observations of the group in the field, and to determine how such reportage could be implemented without bogging down the field team. This turns out to be a much more complex set of issues than one might imagine at first glance, and the work that the RMMS put into resolving them was to prove very valuable when the time finally came to implement mission simulations at Flashline Station in 2001. But what turned out to be of critical importance in the spring of 2000 was that in the course of doing these mountain field exercises, the RMMS had developed a kind of esprit de corps and was capable of mobilizing its members as a team.

Because as events transpired, they would be needed.

THE VOLUNTEERS MOBILIZE

John Kunz was an excellent craftsman, and the work he and the Infracomp team did was of high quality. But by early June it had become all too clear that he just didn't have the manpower to get the job done on time. I hired some extra workers from a company called Mesa Fiberglass to help him, but that wasn't enough. I sent down some of my own employees from Pioneer Astronautics, but that still was not sufficient. So I put out a call for volunteers to the RMMS.

The RMMS answered the call. More than a dozen of them came down to work in the factory, some, such as Dewey Anderson, for up to two weeks straight at no pay. Some did muscle work, some did grunge work, and some did skilled work.

One of the highly skilled volunteers was a young woman named Emily Burrows, a graphics designer from Boulder. It was Emily who created and laid out all the logos and other graphics that adorned the hab. But Emily was more than a graphics designer. She was also, by any measure, a strikingly attractive woman, and she had three or four admirers who would do pretty much whatever she told them to do. These were all husky lads, and we got a lot of labor out of them. Hey, it worked.

While this was going on, Kurt Micheels was assembling all the tools, including a trailer and an assemblable crane, that would be needed to construct the station once it was landed on Devon. Frank Schubert, a Mars Society member who was a homebuilder by trade, had also stepped forward, and together with some of his employees and a few RMMS volunteers, came down to the factory to knock out a set of interior partitions to create the staterooms (or "bunks") for the station.

On the evening of June 26, there was still much left to do. Ship-out was scheduled for the next day. So a pack of RMMS volunteers stayed late to work at the factory. It was raining that night, and much of the stuff that needed to be bundled onto trucking palettes was out in the back lot in the open. The volunteers worked through the night and in the rain to get the job done.

The next day, loaded on three large tractor-trailer rigs, Flashline Station shipped out for the West Coast. The trucks reached Moffett Field three days later, handing over the cargo to the U.S. Marine Corps 4th Air Delivery Battalion. On July 1, the first of the C-130 flights took components of the station and headed north.

It was anticipated that the station, its interior partitions, the crane, trailer, and other tools would require close to the complete load of three Hercs. Pascal said the HMP would have just a small bit of its own supplies to add in, and therefore proposed that the cost of the flights be split 90 percent for the Mars Society and 10 percent for the HMP. Given the indicated split in the planeloads, that seemed reasonable. In the actual event, however, the HMP showed up with a complete Herc's worth of cargo, and had someone at Moffett Field who directed the Marines to carry it in place of the station.

This threatened to derail the entire project. But the Marines saved the day by scheduling a fourth Herc flight on the spot. By July 3, the primary components of the station, along with Kurt Micheels, John Kunz, and a paid team of construction workers, were all in Resolute Bay, ready to hit Devon Island.

On July 4, the construction team climbed into Twin Otters and made their transit to Devon. The next day, the paradrops would begin.

Military cargo paradrops in general only have about a 90 percent success rate. The military deals with this fact easily by dropping in twice what they need. But we did not have this luxury—we had only one hab. We needed to bat a thousand. This would take both skill and luck. The delivery of the station would require five C-130 sorties from Resolute Bay, with a total of seven drops to be carried out. Winning seven out of seven times on a 90 percent shot calculates out to a success probability of 48 percent. We crossed our fingers.

The weather on July 5 included low ceilings and gusty winds, but the Marines were up for it. Three sorties from Resolute to Devon were flown, implementing five of the seven drops. These carried the walls, legs, and some of the dome sections of the station. Despite the adverse weather conditions, these drops were largely successful: the payloads were delivered to the ground safely. There was a minor problem in that the wind blew the parachutes pretty far away from the Haynes Ridge target construction site, with many of them landing in the crater. But that did not concern us greatly—after all, we had considered this possibility too and were bringing in a trailer to deal with it. So with five of the seven drops safely behind us, we were overjoyed. We put out a press release trumpeting our success. The Marines flew the Herc back to Moffett Field to pick up the last load of cargo. The two final drops were scheduled for Saturday, July 8.

I was still in Denver on July 8. My firm was working on Small Business Innovative Research (SBIR) proposals that were due on the fourteenth, and the plan was that I would travel to the Arctic immediately after that date. So on Saturday, July 8, I went to the office to work on proposals and wait for the phone call or e-mail that would confirm that the last two drops had been successful. It did not come. I e-mailed Pascal, Kurt, and Marc Boucher in the Arctic to find out what was going on. I received no answer.

On Sunday, I went into the office again to continue the vigil. Finally, around noon, I received a call by satellite phone from Pascal Lee.

The news was bad.

6. CRISIS IN THE ARCTIC

The final paradrop had failed.

At an altitude of 1000 feet, the payload had separated from the parachute, and then fallen at high speed to crash into Haynes Ridge. The crane, the trailer, and the fiberglass floors for the habitat were completely destroyed in the impact.

Pascal, Kurt, and Marc Boucher were all together in the satellite phone tent. The situation was hopeless, Kurt explained. There were no floors for the hab. Without the trailer, the 800-pound, 20-foot-long fiberglass wall sections that had fallen into the crater could not be moved to the construction site. Even if somehow they were, without a crane they could not be lifted into position, nor could the twelve 350-pound dome sections be placed on top of them. The construction team was unanimous in declaring that building the hab without the crane was impossible. They wanted to leave the island.

Marc Boucher urged me to listen to my people on the scene, and face the facts. The hab would not be built this year. The best thing we could do would be to "winterize" the dropped components, "consolidate," and come back next year and try again.

I was not willing to give up so easily. There are other ways to build things than by using cranes. We could use scaffolding or helicopters to raise the wall panels. We could build replacement floors out of wood. I argued strenuously against quitting five days into the field season.

Kurt would not listen, nor would Boucher. The weather was horrible, and everyone was demoralized. I simply needed to accept reality and understand that construction was not going to happen this year. The only one of the three who seemed at all open to the possibility of going forward was Pascal. He said that from where he stood, things looked pretty bad. But if I could devise an alternative plan, and come up with the means for implementing it, he would be willing to entertain the idea of continuing construction. In the meantime, his focus was going to be on the cleanup effort. The smashed floors had scattered fiberglass insulation all over the place. If it was not picked up, the Nunavut environmental regulators might shut down the HMP. On that note, the telecom ended.

I immediately hit the telephone and started calling such contacts as I had in Resolute Bay, explaining the situation. Someone at Polar Shelf directed me to what sounded like "Ozzie Courage" of "Courage Enterprises," with "Courage" pronounced the French way, cour—raje.

Ozzie Courage turned out to be Aziz Kheraj, the owner of the Resolute's South Camp Inn, and of several other businesses, including a construction company and a metal shop. Yes, he said, construction could be done with scaffolding or with helos. He had scaffolding he could rent us, and he gave some contacts who had helicopters. He could provide us with wood for the floors, and could probably fabricate a small trailer in his shop. If we sent him the crane parts, he might be able to fix them. Alternatively, he could fabricate a large tripod. Together with block and tackle he could supply, this could serve as a replacement for the crane.

Following Aziz's lead, I spoke to Graydon Kowal, the owner of Guardian Helicopter, who had two helicopters on the island, supporting operations at the Noranda mine, just 85 miles away from Haughton Crater. The helos were an A-star (1400-pound lift capability) and a 206 (900-pound lift capability). Either one was capable of lifting the dispersed wall panels and domes and moving them to the construction site. The more powerful A-star could also be used to lift walls and domes into place during construction. He would be happy to rent the A-star to us for $730 US per hour, plus gas, or the 206 for $530 per hour, plus gas. The helos ran on the same kind of aviation fuel the Twin Otters did, which was good because there was a stockpile of this available at the HMP camp. It would take only a few hours for either helo to move the habitat parts from the crater to the construction site. Then, once we were ready for construction, the A-star could return and lift the panels and domes. We could have the walls up in one day and the domes in another. He gave me the name and number of Alan Huard, the manager of the Noranda mine, to arrange the details. I spoke to Mr. Huard, and he was happy to help. The helicopters were being used at Noranda intermittently. Pascal should just give him a call and he would arrange to have one sent over on a day it was not needed.

I also had my guys at Pioneer start looking for a replacement modular crane. In due course, one was located in Manitoba. Somewhat larger than our original crane, it was available and could be packed up and shipped to Resolute, with an ETA of ten days.

At this time, Frank Schubert, the Mars Society homebuilder who had led in the crafting of the interior wall sections of the station, was in Yellowknife. He had been sent north in a second wave following Kurt and the construction crew, because it was thought that his skills in conducting the interior build-out would not be needed until later in the project, when the basic hab shell was already up. Schubert should have reached Resolute Bay on Saturday (the day before), but as a result of the usual snafus of Arctic travel had been stranded in Yellowknife just as Kurt and I had been the previous year. I called up Frank to tell him what was going on. He was a homebuilder, what did he think?

Frank agreed with me: there was more than one way to skin a cat. He had built houses on the sides of cliffs, and done other "oddball projects."

When you were doing stuff like that, you needed to be ready to change your plans, because the first ideas you had rarely worked. The crane had never impressed him as a smart way to go anyway. If it were up to him, he would try scaffolding. I told him about Aziz and the resources he was offering, and recommended that he look him up once he reached Resolute Bay. Frank said he would do so.

By 4 P.M. on Sunday, July 9, I had located these resources, and quickly e-mailed the news to the group on Devon. Pascal did not answer. Kurt and Marc Boucher answered, but could only see problems. A heated e-mail exchange rapidly developed between me and these two.

On the evening of July 10, for example, the following interchange occurred between Kurt and me. For clarity, I have put Kurt's original remarks in plain type, and my responses in italics.

> Robert . . .
>
> This is to advise you of current developments. The construction team and I met with Pascal, Marc, and Andy [Liebman, of the Discovery Channel]. We came to the following conclusions:
>
> 1. It will take too much time to find and move a crane here. The construction team is not free to remain an additional 2–3 weeks. Dale [Dale Cameron, the leader of the construction team] concurred with this. We would like to research the assembly problem in detail and arrive at a better/safer solution after our return.
>
>   *There are other ways to construct without a crane*
>   **helo*
>   **block and tackle + tripod*
>   **brute force + scaffolding*
>   *etc.*
>
> 2. The helicopter was deemed to be undesirable and unsafe. No one here is familiar with these operations. Further, we could damage

the panels considering the wind. And . . . it has been too windy for helo operations since we have been here. The current weather report provides no improvement.

We can bring in people who are familiar with helo ops.

3. We have no way to move the panels at present, but would like the option to look for a better, lighter trailer and more powerful 4 wheel drive ATV.

   That's bullshit. Helos are available that can move everything easily right to the site.

4. We will winterize everything here before we leave. We will move the existing crane to the landing site and begin shipping it to Resolute. Do you want the tools returned to the US or stored in Resolute?

5. I will prepare a schedule for next year showing an extended field season. It will allow for assembly, and, perhaps, as much as 5–6 weeks of hab occupation. Andy is positive about this and can provide this to Discovery.

   If you give up after one week this year, there is no reason to believe you will accomplish anything next year.

   We need to build the hab this year.

   Get the wood for the decks, and start making them. Get a helo and move everything to the construction site. Start preparing the panels for assembly.

   I'll be there on Saturday. Please don't waste the week.

   Robert

In a series of messages, which I also circulated to the Mars Society Steering Committee members who were on the Flashline management team, I tried to make clear to Marc, Kurt, and Pascal the need to move forward this year. We had the full team in place, including John Kunz, who we could probably not get to the Arctic again. The availability of two helos just thirty flight minutes away was a godsend. We had a full field season in front of us, and the weather would be no better next year. If we gave up now, the credibility of the Mars Society would be destroyed.

Mars Society Canada President Marc Boucher disagreed. In a message sent early on the morning of Tuesday, July 11, he said, "I do not agree with Robert that if we do not build the habitat this year that the project would be dead. Such a statement is ridiculous."

To this I replied, "No, it's true. We're there now with a team and lots of equipment, with a 5 week season before us. If we can't do it now, we can't do it."

Later that day, Kurt sent word that he was pulling the construction team out Saturday, July 15. This prompted an angry response from me, and several other Steering Committee members jumped into the fray.

My wife Maggie wrote, "The idea of pulling the troops out Saturday is neither fair nor appropriate. It may not have been planned this way, but it appears as though the idea is to get the workers off the island before Robert arrives, to avoid any input from him. It is imperative that no mass departures be made in advance of Robert's arrival. Communication is very difficult. We must wait for calls from Devon, as we cannot get a line to the sat phone. In good faith, there should not be any early pull-out. . . . It is clearly and simply not appropriate to quit."

Chris McKay immediately responded. "I agree with Maggie. Failure is part of risk. Failure to try is not acceptable."

Kurt, however, wasn't having any of it. The construction team had tried to move some of the dome sections that fell on the crater slope up to the ridge, and the activity had terrified him. "The incident that occurred here could have caused the loss of the Marine aircraft. There were numerous times when manhauling dome panels out of the crater that someone stumbled and nearly fell under the panel onto jagged rocks. It is truly amazing we have not killed anyone yet.

"As the project manager on scene, stopping the project, for now, seems the best thing to do. Please respect this extremely well informed opinion."

Carol Stoker tried to respond to this with sweet reason. She wrote: "I think the future of the Mars Society, and certainly the future of the Mars Base analog project (the FMARS) IS at stake. To turn back and not make every reasonable effort to recover from this setback would be to show the world that the Mars Society does not have what it takes to do hard things. It is at times of adversity that the truly committed show their true grit. Working in the polar regions is not easy, and often not fun. To use the Antarctica motto: 'It's a harsh continent.' If we don't continue to move forward in the face of this setback, many will say 'they didn't have what it takes.' I watched this happen after the failed attempt to put a robot down on Mount Erebus. The team never recovered from that failure in spite of later successes. They became a public joke. Once faith is lost in us, we will no longer be able to command the respect of sponsors. We will not be able to raise the money to 'come back next year and try again.' We will not be able to raise our own spirits to do it."

It was good that the Steering Committee got in their input on Tuesday, because that night I finally managed to reach Pascal. With the HMP leader not answering my e-mails, the only way to reach him was by satellite phone. This was difficult, because the phone was kept in a tent on a small hill about a hundred meters away from the HMP base camp. Generally speaking, there was no one there. On one occasion, I had gotten through to Marc Boucher, but he refused to get Pascal. On Tuesday night, Boucher picked up the phone again. When I asked to speak to Pascal, Boucher replied, "What good would that do?" But Pascal was right there, and seeing that the call was for him, insisted on taking it.

The situation was worse that I had thought. Rather than continuing any attempts to move hab components to the construction site as ordered, the construction team had decided to return to Resolute Bay "for some R&R." Kurt was going too, and wanted to move the tools out with them.

I told Pascal that under no circumstances should the tools be flown back to Resolute. If they were, the construction team would put them on the first plane south to preclude any effort by anyone else to build the station. Pascal agreed to prevent the tools from being removed. I asked him to call Alan Huard and get a helo in to move the hab materials to the construction site. He refused this request, saying that he had a helo of his own coming in on July 21, and if we wanted to use a helo, we would have to use that one. This seemed crazy, because it would waste ten precious days, but he wouldn't budge. So I asked him to muster labor, and move the hab components to the construction site. The HMP had ATVs and an ATV trailer. Using these, the hab's legs and dome sections could at least be moved. Pascal refused this as well, saying that "it hadn't been decided yet" that we would do construction this year, so the value of moving everything wasn't clear. I would be up on Saturday, and we could talk about what to do then. What he did say, however, was that if I could convince him that I had a credible plan, and if the success of the plan depended on it, that he would be willing to keep the HMP base camp open through August 15 to support construction operations. The field season was scheduled to end August 5. This was a significant concession and I thanked him for it.

Pascal seemed to be of divided mind about the project. His primary concern was for the HMP. For him, the success or failure of the Flashline Station project was of secondary importance, and some people were warning him that if we had an accidental injury, it could put the HMP in danger. Yet, although he had not done much for the Flashline Station program, he had provided the initiating suggestion, and as Project Scientist had his personal prestige on the line. The HMP also stood to make money by charging logistics fees to the press who came north to cover the Flashline Station. Moreover, some twenty journalists from around the world were already at his camp, and if he gave up, they would cover that too. Finally, Chris McKay, who controlled his NASA funding, had made it clear he wanted us to push forward. All this left Pascal in a fundamentally ambivalent position. Unlike the construction team, he was not freaked out by the cold, and was not looking for an excuse to get on the first plane south from the Arctic.

But his divided commitments appeared to render him incapable of positive action.

Kurt and the construction team flew out of Devon that night, leaving the tools behind them, but no work under way on the island. It was imperative to try to get things moving. So on Wednesday, July 12, Aziz flew in to inspect the crane, assess the general construction situation, and talk with Pascal about getting work started. Pascal, however, would not talk with him. On Friday the fourteenth, Frank Schubert, who had spent most of the previous week staying at Aziz's South Camp Inn developing a construction plan in consultation with the Arctic contractor, flew into Devon to try to explain the plan to Pascal. Pascal had him immediately thrown off the island.

On Friday, July 14, as I drove to the Denver airport I received a phone call in my car from a journalist.

"So, Dr. Zubrin," he said. "How would you compare the failure of your program with that of the Mars Polar Lander?" He was referring to an unmanned probe that was destroyed by crashing into Mars in December 1999.

"There's a similarity in that we both hit a rock," I replied. "But the difference is that we have a human crew on the scene and we are going to find a way out of this."

"So," he said, paraphrasing Apollo 13 Mission Director Gene Kranz, "you're saying 'failure is not an option'?"

"Yeah, that's right. Failure is not an option."

On that note, I headed north. I reached Resolute Bay the next day.

I was met at the Resolute airport by Aziz and Frank, and very shortly afterward, Kurt Micheels. He approached me and gave me a letter, which said that the team had decided that no construction of the hab could occur this year, and that if I insisted on hiring Aziz Kheraj, and proceeding with the job, he resigned. I promptly accepted his resignation. There was really no choice. I was intent on proceeding with construction, he did not believe it could be done.

I then went with Frank and Aziz back to Aziz's hotel briefly to confer with them, and they showed me their rather detailed plans for construction of the habitat using scaffolding. The plans looked good to me, but the key question was putting together a crew that could implement them. Of the old construction team, the only one we really needed was John Kunz. Kunz had created the fiberglass structure and knew it inside and out. He knew where he had inserted hard points to which attachments could be made, and he knew the special art of making repairs to the material.

So Frank and I went to the Narwhal hotel, where the other group was staying. I managed to get Kunz alone and talk with him for about half an hour, making clear my desire to continue and keep him on board. This was one of the tensest conversations I have ever had. If Kunz decided to call it quits, we would really be in trouble. I reminded him of the visions he and his father had for their technology, and pointed out how the Mars Arctic Research Station would showcase it before the world. I pointed out how I had stuck by him; when he could not deliver the hab on time, I had hired additional workers to get it done at no cost to him. Now I needed him to stick by me. Kunz appeared to be mixed in his emotions. All the other guys on the construction team were pressuring him to leave along with them. The last thing they wanted was to see the hab built after they had fled from the job, as this would expose them as a bunch of chickens. But Kunz had a lot more at stake. He told me he would think about it. He, Kurt, Dale, and some of the others had left their stuff behind on Devon and were going back to get it. He would meet me on the island and give me my answer then.

So with that uncomfortable ambiguity in the background, Frank and I flew on to Devon, along with NASA Ames scientists Kelly Snook and Bill Clancey. In an attempt to develop some degree of oversight of HMP operations, NASA management had appointed Kelly Snook to be Program Manager of the HMP. This made her technically Pascal's boss, as he was supposed to be serving in the capacity of Principal Investigator under her direction. In practice, however, Pascal continued to run the HMP camp as he pleased, while sending Kelly back and forth on errands between Devon and Resolute.

The HMP camp this year was not located in the crater. Instead it was on a little plateau just to the west of an interesting rock formation that everyone called The Fortress, which itself marked the western terminus of a long band of flat dirt that served as the airstrip for the Twin Otters supplying the camp. To the south of the airstrip was a valley, through which ran the small river we called the Lowell Canal, and across the Canal, rising above it about 300 feet in elevation was rock-covered Haynes Ridge, on which we hoped to build the station. Just south of Haynes Ridge was the crater, into which most of the station parts had fallen. The distance from the HMP camp to the proposed Flashline Station building site was about 2 kilometers. With respect to the circular crater as a whole, it was placed northwest, roughly in the 11 o'clock position, with 12 representing due north.

After landing, Frank and I pitched our tents, and then, in the middle of the cold but sunlit night, met in the HMP's mess tent with Pascal and Marc Boucher. Frank explained the plan that he and Aziz had developed for constructing the hab using scaffolding. It was important to win Pascal over, not because, as he appeared to believe, he could have stopped us with a command. Had Pascal refused, we would have continued regardless. But we would have had to do it without the support of the resources of the HMP camp. This would have made things much harder.

During the meeting, Boucher did not say much. Pascal, however, seemed fairly favorably impressed with the plan, and with Frank Schubert. But he said that he would only be willing to support proceeding if John Kunz would agree to stick around.

It was afternoon the next day before the Twin Otter carrying Kunz and Kurt returned, bringing with it Dale Cameron and some of the other construction workers. Frank and I each took another turn at talking with Kunz, who remained enigmatic. Then we filed into one of the science tents for a tense council of war that would resolve the situation. Attending were Frank, Pascal, Kurt, Marc Boucher, Kunz, Dale Cameron, Mesa fiberglass worker Chester Snyder, and I.

Frank started explaining his plan. Kunz nodded agreement, indicating that he thought it might be feasible. I argued for the plan, Kurt against. Boucher kept quiet. Dale Cameron said the plan as written wouldn't work, but it might if we could get shoring scaffolding (heavy-duty) in place of the ordinary scaffolding that we actually had. That shifted the conversation into how-to technical thinking, and it gradually became clear that there was a consensus among the engineering and construction types in the room that the job could be done. Pascal moved toward the consensus. "So now we have a plan," he said.

Cameron said, "Yeah, it's all right by me. But as far as I am concerned, Kurt is construction manager."

That was unacceptable. Kurt had made it plain that he thought we could not build the hab this year, and that he believed Frank's plan was far too dangerous to attempt. Kurt had done a very good job as an architect, but his construction experience was limited to big jobs done according to plan by large organizations operating under civilized conditions. He was a careful planner, but the situation called for an improviser. Kurt was not a coward. The risks he had identified in our plan were real. But there was a fundamental philosophical difference between us. He was not willing to take the risks necessary to build the station, and I was.

Furthermore, he had already resigned. So I intervened forcefully. "That's out of the question," I said. "He's being replaced with Frank Schubert."

Dale quit on the spot and walked out of the tent. That night, he and Kurt flew back to Resolute. The old team was gone. Now we had to create a new one.

Transferring leadership of the construction job to Frank Schubert was probably the most important decision I made on Devon Island that summer. A word about it is in order.

That the construction job could not continue under the leadership of any member of the old team was clear. But why give the responsibility to Frank? On the surface he was not an obvious choice.

Frank Schubert was a construction man, yes. But before he had entered construction, he had been a rock musician associated with the pioneering New Wave band Devo. Aged fifty, he still had some of the aspect of a hippie rocker about him, and he was given to telling stories about various famous movie stars that he had supposedly once dated. He had also adopted a few miscellaneous New Age ideas concerning such things as hidden energy in the ground. All of this made him seem a little odd. But in the discussions we had, Frank had shown himself not only very knowledgeable about construction engineering but incredibly creative as well. Where others could see only problems, he saw possibilities. He was the sort of person who could come up with a dozen new plans before breakfast, and gleefully revise or abandon them all when the touch of a tool told him they were not going to work. I felt very strongly that this was the sort of attitude we would need. Flashline Station would not be built according to a preconceived plan. We were going to have to wing it every step of the way, and the construction leader would have to be someone comfortable operating that way. Frank also had a puckish sense of humor, which would come in handy in getting past the setbacks we would be sure to encounter.

I knew that Frank Schubert was kind of wild. I could have hired a more standard-issue Arctic construction leader through Aziz. But the Flashline construction project was a wild horse. It needed a wild rider. So I chose Frank.

THE KUNZMOBILE

On Monday we flew in parts of a wrecked baggage cart that Kunz had found at Resolute Airport. The parts had consisted of just the axles and wheels, with a few metal pieces hanging on. Kunz figured that they could be linked together by some two-by-fours with some wooden cross bracing to create a crude trailer. So Monday morning, we hammered the thing together. It was as ugly as all get-out, but you could attach it to the back of an ATV and it would roll. We dubbed it the Kunzmobile.

Shortly after lunch I went down to camp and told everyone we had a trailer and needed volunteers to help us haul walls. There were Mars Society members among the HMP scientists, and they stepped forward; Singapore-born Canadian Darlene Lim of the University of Toronto, Bulgaria-born Canadian Margarita Marinova of MIT, Samson Inutuvak, an Inuk studying physics at the University of Halifax, Lieutenant Colonel George Martin, an African-American Air Force doctor working with the astronaut corps at NASA JSC. Several members of the Japanese NHK TV crew also volunteered. Together with Frank, Kunz, Chester, and me, this wonderfully mixed set of representatives of the various peoples of the Earth formed our first work crew.

We marched over to the ridge and then down into the crater, where a group of six wall panels were lying. As the ridge right near the construction site was too steep to drive an ATV down, let alone one towing the rickety Kunzmobile, Frank had to drive it down from the ridge by a long circumferential route to the east. Then, lifting together, we heaved the 800-pound panel onto the makeshift trailer. The Kunzmobile groaned but it did not break.

So we strapped the panel down, and with the whole team pushing to help the ATV move the heavy load over the variously muddy or rocky ground, we set off. As we hauled the wall panel over a half-kilometer circuit that took it to the east and then up and out of the crater, someone on the team started to whistle the "Colonel Bogie March," the military tune made famous in the movie *The Bridge on the River Kwai*. This was immediately taken up by the whole team, which was interesting since two of those helping to push the panel-laden trailer up the crater slope were Japanese. Then, reaching the summit, we pushed back west cross the ridge to reached the construction site. We had done it! We had moved one panel. The team was elated.

We went down into the crater for another. As we loaded it, a light rain began to descend. It wasn't much, just a drizzle. But the temperature was near zero, and the wind was up, and the cold wetness gradually penetrated our clothes to deliver a deep chill. But still, we moved a second panel, then a third, a fourth through the freezing rain. By the time we hauled out the fifth, it was dinnertime. But we were in no mood to stop. There was one panel to go. It cost us our hot meal that night, but we got the sixth panel too. We got back to camp that evening chilled to the bone but with morale sky-high. A member of the old construction team had predicted it would take two weeks just to move the hab components to the building site. We had moved half the wall segments in a single day. It was spectacular.

Monday's success caused a buzz in the camp, and Tuesday morning we had a lot of volunteers. Not only did these include Mars Society members, but even people like Marianne Douglas, Darlene Lim's thesis adviser at the University of Toronto, turned out to pitch in with manual labor. Unfortunately, the front section of the Kunzmobile fell apart, preventing us from moving any wall sections that morning. So while Frank and Kunz tried to fix it, I took the volunteers down into the crater, and using an ATV and an ATV cart trailer, we managed to retrieve three of the 400-pound steel primary legs of the habitat.

Then it was afternoon, and some of the scientist volunteers left to do their fieldwork, but their places were filled by a group of Inuit high school boys that Pascal had the Mayor of Resolute find for us to hire. These were Greg Kalluk, Dennis Idlout, Joannie Pudluk, Enooki Idlout, and Jeff Amarualik. They were to prove hard workers and extremely bright. They were, however, first-generation children of hunter-gatherers, and as such hailed from a completely different world from the scientists and Ph.D. students who constituted the other primary element of the work crew. During the same winter months the scientists would spend writing grant proposals, these kids would be hunting seals to feed their dogs. The way these two groups came together to form a team that clicked was one of the marvels of that summer on Devon.

In addition to the Inuit youngsters, an adult Inuk also joined the effort. This was Joe Amarualik. A ranger from Resolute Bay with seventeen polar bears to his credit, Joe carried a Lee-Enfield .303, but was equally handy with tools. The fact that Joe toted a rifle was quite welcome, since as a result of a screwup, the HMP had failed to bring any shotguns across the border that year. Thus the only shotguns present at the HMP camp were the 12-gauge Mosberg pump-action street sweepers packed by Canadians Darlene Lim and Marianne Douglas. This placed the large majority of available firepower firmly in female hands—which can be a terrifying thought if you are a male. Actually, I did not mind the fact of local feminine military superiority; Darlene and Marianne are both very nice people, just the sort who should control the world's armaments. The problem was that as working field scientists, Darlene and Marianne were only intermittently present at the Flashline work site. This left us without polar bear protection most of the time. Joe's addition to the team remedied this weakness. He also acted as a kind of guide to the Inuit boys who were on the job, several of whom were his relations.

The broken front axle of the Kunzmobile proved unfixable with the resources available on the island, so Chester Snyder said he would take it back to Resolute and see if he could get it fixed in Aziz's shop. He never returned. But Frank developed a workaround in which he substituted an ATV cart for the Kunzmobile's front axle, attaching it with a set of bolts to the long two-by-fours that emanated forward from the rusty metal parts connected to the old baggage cart's rear axle. Using this truly makeshift trailer and our combined Inuit/scientist work team, we managed to transport four of the remaining wall panels from the second drop zone.

Of all the payloads, the domes had been paradropped the most accurately. This had allowed the old construction team to move a few of the closest dome sections to the site. On Tuesday, Frank and some volunteers moved the rest. This left us with only two wall panels and three legs left to be transported. In the late afternoon, Frank, Joe, Jeff, and I also bolted two pairs of wall panels together with a leg for each pair, forming the first two of the six "trios" that would be needed to build the hab.

A minor crisis broke out Tuesday evening, when John Kunz announced he wanted to go to Resolute "to call his wife." This concerned me greatly, as we needed Kunz. Some members of the old construction crew were still in Resolute, and had apparently been successful in convincing Chester Snyder not to come back. We couldn't afford to have Kunz go the same way. He said he needed to go to Resolute because the quality of reception on Pascal's satellite phone was too poor to have a decent conversation. Discovery Channel had an Internet phone attached to its satellite communication setup, and it had fairly good sound quality. Andy Liebman was more than happy to make it available. I convinced Kunz to use this system and he stayed.

I woke up Wednesday morning feeling sick. My down ski jacket had been soaked through on Monday, rendering it useless and subjecting me to a severe chill. I had tried to correct this on Tuesday by discarding it in favor of a Gore-Tex military shell jacket with sweaters underneath, and woolen Swedish army pants plus thermal underwear for my legs. These were good choices, but Monday's chill had already done its work. I had contracted a horrible cough, which was to plague me for the rest of the season, especially at night, when it made sleep difficult or impossible.

Nevertheless, in the course of the day we managed to cart all the remaining wall panels and legs to the construction site. Despite various breakdowns of the Kunzmobile, we were able to patch it up each time, in one way or another, and get the objects there. The transporting job was done. In three days of manic work, we had accomplished what the old crew had predicted would take two weeks.

We then proceeded to bolt together three more pairs of wall panels, and each of those pairs to a leg, leaving only two wall panels and a leg unattached.

As the day wore on, the weather started to improve, and by late afternoon the high winds and freezing rain of the previous days were replaced with light breezes, blue skies, and temperature in the 40°F range. This is about as good as it gets on Devon Island, and it set the stage for the most critical decision of the entire construction effort.

Did we dare to try to raise the walls of the hab with the equipment available?

Until now, we had been engaged in hard work hauling heavy objects over difficult ground, but nothing particularly dangerous. But the heavy wall panels were 20 feet long by 7 feet wide, and once raised could catch the wind to act as powerful sails that would break free of anyone trying to control them from the ground and come crashing down to cause injury or death to those beneath. The decision taken the previous Sunday to initiate construction had only covered the nonhazardous transportation phase. Raising the walls was the scary part. Frank had a plan, and around 11 P.M. on the sunlit night of July 19, a meeting was held in the HMP mess tent to consider it. The meeting was attended by me, Frank Schubert, John Kunz, Pascal Lee, Marc Boucher, John Schutt, John Schutt's wife, A.C., an able carpenter, and Don Gunderson, a mechanic whom Aziz had hired to help us.

As he had gained knowledge of local conditions since Sunday, Frank had changed his plan several times. Now he had an idea of how to get the walls up without even using a scaffold. He laid it out as follows:

The hab was comprised of twelve wall panels, twelve dome panels, a hub that joined all the dome panels at the top, and six legs. We had already joined ten of the walls and five of the legs to form five of the six "trios" (composed of two walls and leg each) that themselves would be joined to form the primary structure. Weighing 2000 pounds, and measuring 24 feet from leg-bottom to wall-section top, each trio would be pushed up from behind by work crews with bracing timbers. Then a cable would be run from the top of the wall to a piece of the crane that had survived the crash, which we anchored into the ground, where it was attached to a winch. We would use the winch to pull the trio up, levering it against the ground by its own leg, while teams of workers operating on its side stabilized it with guylines. Once the trio was erect, its leg would be anchored into the ground, and its guylines moored in place to hold it vertical in the manner of a mast. Then another trio would be lifted, anchored, and moored, and the counterclockwise edge of one of its wall panels attached to the clockwise edge of the trio that had been raised before it. Proceeding around in a circle, the cylindrical shell of the station would thus be erected by raising six trios in succession.

The plan would be incredibly dangerous if there was any wind, as the massive trios had a double panel sail area that would make them impossible to control. But the wind had dropped to near zero. The meeting approved the plan. Upon leaving it, a little after midnight, Frank and I had a private talk. Since everything depended upon weather, we needed to push like mad to get the hab up before conditions turned bad again. Furthermore, while Pascal had in principle made an offer to keep the HMP camp open for an extended season to enable construction, neither Frank nor I could actually stay in the Arctic beyond July 29, and it was evident from the history of the past two weeks that if we were not there, no work would be done. So the hab work team would not attend morning briefings, or break for lunch or dinner. We would bring dried food from camp, eat our meals on site, and push full speed ahead with fifteen-hour workdays. There was no other way. We were in a race against time.

WALLS RISING FROM THE DEAD

Thursday, July 20, dawned clear and bright, with wind conditions practically dead calm. As Frank and I had agreed, the work team cleared camp immediately after breakfast, without wasting time listening to Pascal's customary briefing about HMP ATV deployments. We got to the site and began preparations to lift the first panel. Around 11 A.M., when we were almost ready to begin, a Twin Otter landed carrying with it several passengers. Instead of going to camp, one of the passengers immediately walked across the valley and presented himself at the work site.

The new arrival's name was Matt Smola, the foreman of Frank's construction outfit back in Denver. After taking over as project manager from Kurt on Sunday, Frank had requested that Matt be flown up to assist him. He had made it to the Arctic from Denver in record time by flying to Ottawa and then coming up by the eastern route, which included a midweek flight to Resolute Bay out of the Nunavut capital of Iqaluit. A husky blue-collar guy of slightly more than average height, Matt was incredibly strong. He was also great with tools and good at teaching unskilled volunteers how to use them. His ability to come up with creative expedients to allow the job to move forward after existing ideas failed was matched only by Frank's. He was to prove critical to our success.

So with all hands pushing, we levered the trio up to an angle of about 45 degrees against the ground. Then, with some of us holding guylines on the right and some on the left, John Kunz turned on the little power winch and the massive trio rose like a fallen man rising from the dead. We cheered as it rose, and then as soon as it was up, we ran to stake our guylines down while Frank and Matt pounded stakes into the holes in the foot of its massive steel leg.

Then there it stood proudly upright, the first trio, containing the hab's front airlock door with the Mars Society's and Discovery's logos emblazoned on it. Marc Boucher snapped a picture of it and dashed away to camp on an ATV to send it off over the Internet. *Space News* was running a story on the failure of our project. They wanted a photo of the crash wreckage. He sent them that one instead.

One trio was up, but there were five more to go. We spent the afternoon preparing the second one for lifting. Shortly before we were ready to raise it, Pascal showed up, escorting a party of some twenty journalists out to the work site from camp. He had me give them an explanation of the state of the project, which I had to kind of croak out, since my cough was affecting my ability to speak. Then he walked them over to the edge of the crater and started giving them all a lecture on its unique geology.

Well, now it was time to lift the second trio, and I didn't care whether the press watched us do it or not. So we pushed it up into the 45-degree position, got to our guyline stations, and started up the little winch. This time as it rose, however, we noticed that Frank's guyline was crossed with one of the lines holding up the first panel. Frank detached his guyline from its hook in the ground and tried to run around and uncross the two lines. But when he did so, he momentarily loosened tension on his line and we lost control of the trio. It came toppling down, smashing into one of the other trios that were lying on the ground nearby. The event was so tumultuous that it really shocked everybody, and we thought we had suffered major damage. But Kunz's wonder material saved the day. Incredibly, the only damage was a cracked flange on the edge of one of the wall panels. This was something that fiberglass craftsman Kunz could easily repair.

The event had a comical side, however. Twenty journalists had been standing not fifty yards away from the construction site at the time of the panel crash, but none had seen it because they were all facing the crater as they listened to Pascal's geology lecture. Observing the journalists obliviousness in the moments immediately after the crash, the members of the work crew looked at one another and burst out laughing.

The crash of the second trio stopped work for the evening while the repair resin in the wounded flange set. So, contrary to plan, we did get back to camp in time for dinner that night. Afterward I had my first chance in days to go into the com tent and read my e-mail and check out what was happening on the Net.

There, on the front page of CNN.com and MSNBC.com was a photograph. It was the shot taken earlier by Marc Boucher, showing our wall-panel trio standing bravely on its one leg, like a gesture of the defiance of the human spirit against the gods of adversity.

I walked outside and looked over toward the construction site. You could see the trio from camp, two kilometers away, standing tall and white against the dull blue sky of evening.

It was only about 10 P.M., but I was really tired from work, emotional stress, and lack of sleep the night before. So I shuffled across camp to retire to my tent. As I did so, I noticed a group of Mars exobiology graduate students and Inuit boys playing Frisbee together.

MATT AND FRANK'S MIDNIGHT SORTIE

Matt and Frank did not sleep that night, or at least not much. Matt did not believe in the trio construction technique, and he had been vocal about his views. Trying to lift 2000 pounds at a time was just too hard, he said.

Our success with the first trio had appeared to refute him, but our failure with the second had convinced him he was right. Fine, getting one trio up had worked, when there was nothing else to interfere with it. But even if we hadn't made the mistake of crossing lines while raising the second unit, and somehow had gotten it erect, there still would have been the issue of moving it to perfect alignment with the first. Under the circumstances, there was no way for us to do that. So beyond raising the first unit, the trio construction system just wouldn't work.

Matt wanted us to change techniques to one based on lifting just one wall panel at a time. Then we would only have to manage 800 pounds in the air at once, which would be far more tractable than hauling around 2000-pound trios. Exactly how he proposed to do this, however, Matt did not make clear, and Kunz would not hear about changing construction methods unless there was a well-thought-out plan presented on paper.

Matt thought that Kunz's preference for plans and consensus was bullshit. The way to get things done was just to do them. So he woke Frank up around midnight and convinced him to go out to the site so the two of them alone could try out some ideas without bureaucratic interference.

The next morning when I woke up, Matt and Frank were nowhere to be found. They were not at breakfast, and when Kunz, Don, and I arrived at the worksite around 9 A.M., they were not there either.

But a third wall panel had been raised and was attached along its edge to the trio we had set up the day before. The new panel was being supported from below by a crude wooden jig, and held up from above by a line connected to our rickety scaffold, which apparently had been erected during the night.

Kunz was pissed. Without agreement from anyone, Matt and Frank had gone off in the dead of night and secretly implemented an unplanned construction method! Furthermore, they had done it all wrong. They didn't understand the fiberglass structure or how proper attachments were to be made. They had put the wrong bolts through the wrong holes. Kunz stamped his feet. The new panel would have to come down.

Well, if Matt and Frank had spent their night doing this, it was obvious where they were now. I went back to camp and rousted them grumbling from their tents and herded them back to the construction site. When they arrived, Kunz repeated his fit. In response, they just stood there and smiled, leaving me to make the obvious point. "Yes," I said, in as conciliatory a way as I could, "it has to be brought down and put up again correctly. But if using this technique, two men working alone could raise a panel, it's clearly the right way to go."

Don agreed, and Kunz ultimately was forced to do so as well. There was simply no denying that Matt had proven his point.

The new method was this: A jig, which was a wooden object the size of a small bookshelf, would be placed where the bottom of the panel to be raised was supposed to end up. This was directly adjacent to an already raised panel. Then we would haul the aspiring panel into place, with its bottom lying on top of the jig, and its top leaned back to lie within the circumference of the hab cylinder. The scaffold, whose top reached somewhat above the location where the top of the new wall panel would be, was then moved in front of the jig, facing it from the outside of the hab cylinder circle. A line would then be run from the top of the panel to a pulley on top of the scaffold and then down again to the ground on the other side, where our little power winch was anchored. Guylines would be attached to the side of the panel, to keep it under lateral control as it was raised. Then, with four people manning the guy ropes, and two keeping an eye on the bottom of the panel to make sure it didn't slip off the jig, the power winch would be turned on and the panel lifted into position. Then we would bolt it to the nearby panel that was already up and place a high-lift jack under the bottom of the new panel so we could remove the jig. Another panel the next step clockwise around the cylinder would be raised and attached to the previously raised panel in the same way, and then we would raise and bolt on a leg to support the two new panels from the other side. So it would go panel, panel, leg; panel, panel, leg—around the circle until the entire cylinder of wall panels was erected.

As can be seen from the above description, this method of construction really requires more than two people. It had taken hours of hard work involving repeated slips for Matt and Frank to use it to raise one panel alone, as they did not have enough hands to man the guy ropes adequately. But they had done it, and there was no arguing with success.

We adopted this technique and went at it with a will. We worked through lunch and dinner, and by 10 P.M. we had raised four more wall panels. Half of the hab cylinder was now standing.

When we returned to camp, Pascal Lee was waiting for Frank and Matt with a stern lecture.

I had anticipated that this would happen, and briefed the two of them in advance. I did not have much worry about Frank. He was an easygoing guy and would pretend to listen to Pascal while laughing inside. But Matt was another matter. He was a very physical person, and was not the type to take it kindly when subjected to an authoritarian tirade. So as Pascal went on about how they had broken camp procedure by going off at night without his permission, I stood next to Matt. At one point, I saw his fist tighten, so I put my hand on it to suggest restraint, while whispering to him, "It doesn't matter, just ignore him."

Matt, as I said earlier, is a very strong person. I have seen him lift objects weighing 300 pounds. There is not the slightest doubt in my mind that he could have knocked Pascal to pieces in less time than it takes a New Yorker to jump through a subway car door. So I acted to calm Matt down, thereby saving Pascal from a sound thrashing. It was a close call.

V FOR VICTORY

The weather still held fine the next day, Saturday, July 22, and we put up more wall panels with determination. We knew how to do it now, and one panel after another went up like clockwork. You could literally feel the spirit of the team, moving forward like an avalanche. Nothing was going to stop us. Around the time the ninth wall panel went up, the HMP helicopter flew overhead, with someone aiming a video recorder out the side. I suddenly put up my hand with the V for victory sign, and everyone on the job did the same. As the helicopter circled above us, we waved the V signs and cheered. That was the way we felt.

Around 7 P.M., the twelfth wall panel was raised and bolted into place. That was great. There was just one problem though—there was a foot-wide gap between the twelfth panel and the first. As a result of the rocky surface and the crudity of the instruments available, we had constructed the hab out of round.

Kunz was very upset. The hab was his baby, and we had screwed it up with our half-assed building techniques. He went back to camp to try to get some dinner.

But Matt didn't give up so easily. He had us jack the entire hab up off the ground with high lifts, and then he ran a cargo strap all the way around the hab and attached it to a come-along. A come-along is like a little winch for pulling straps tight. Using it, he winched in the cargo strap and *squeezed the entire 12,000-pound cylinder* into its proper circular shape. Just as Kunz returned from his dinner, the trailing flange of panel number 1 slid over the leading flange of panel number 12. Happy as a kid exiting the building on the last day of school, Kunz leaped from his ATV and popped three bolts into the holes linking the two flanges. A half hour later all the bolts were in and tightened, and the hab cylinder had been lowered from the high lifts to stand on the ground on its own six feet.

The walls were up. The cylinder was done. We went back to camp and got drunk on screech rum.

My voice log gives a good sense of the feel of the next few days:

Voice Log, July 22, 2000.

I'm drunk, my legs are shot, my voice is shot, but all the walls of the hab are up. And I feel about as good as a man can feel.

o

Voice Log, morning of Monday, July 24, 2000, Devon Island.

Most of the crew were hung over, and didn't get to work until about one o'clock in the afternoon. There was also a problem with the fact that we only had about a third of the amount of wood we needed to build the decks, which is our next step. Still, we did get to work, and worked from around one in the afternoon until nine at night, and by the end of the day we had built around 80 percent of the first deck, and about 10 percent of the second deck. The weather was fine, and it's fine again today. I've been having some health problems—some laryngitis and a bit of a cold. Last night I was afflicted with a terrible cough. I had to get up at three, walk over to the main tent, and sit there for a while reading and drinking hot Tang, which is a drink I'm coming to enjoy and detest at the same time. But today we hope to finish the decks and start raising the domes. If we get the domes up, that is the last major challenge before the completion of the hab, fitting out the interior. I am resolved that we can do a one-week occupation of the hab, this year.

As an addendum to that, I'll just add for the record that I've just changed all my clothes, with the exception of my heavy wool pants, the Swedish army pants that have proven a godsend on this trip. I have not been able to shower. I did shave yesterday, which made me look a lot better and feel better. I guess it was reflective of the improved morale after the walls of the hab all went together.

o

Voice Log, evening of Monday, July 24, 2000.

Today we got a reasonable start, beginning about nine o'clock, and finished off most of the second deck. We were able to haul up two of the dome panels to the second deck, as well as the pedestal. In the course of doing that we were using the small scaffold to hoist with, and one of the Inuit kids pulled down on the scaffold after everyone else had let go, and pulled it over. It fell, and I was hit by the steel hook on the end of the block and tackle, right in the head. It really hurt, and for a while I was concerned that I may have suffered a concussion, because I did want to go to sleep at that point. But I stayed awake, and while I stopped working for a little while, basically told everybody that I was okay, and kept going.

Earlier I spent most of the day sawing wood to make the joists for the deck. It was actually kind of fun. I have become capable in certain aspects of the carpenter's art, and have been able to make a good contribution to the team in that capacity. But anyway, later in the day I was hit by that steel hook from the falling block and tackle, and I kind of wondered, you know, what winning through on a job like this is worth. So I've been thinking about what the trade here is. My health hasn't been that good. My voice has been terrible. I've been having gummy eyes from all the dust. I've kept it all hidden. But I guess something is making it clear to me that there is a price to be paid for doing stuff like this.

Anyway, at six o'clock John Kunz started acting moody, making all kinds of complaints about the way things were done. The bottom line was, he wanted to break for dinner, and not keep working into the evenings as we had on every previous night. So we did break for dinner, and it's the first hot dinner we've had in about a week or so. But I think we are now actually ready to put up the first two dome panels, and I think it's best that the team be completely fresh when that happens, and not be worn out from working into the evenings. This is the last complicated thing we need to do.

o

Voice Log, Morning of Wednesday, July 26, 2000, Devon Island. Yesterday we worked very hard, and managed to put six of the dome segments of the hab up. We get what help we can at the hab. Typically only about two of the Inuit are available to help us, sometimes none. It got us pretty mad at Pascal, who has assigned them to other duties. But we get some help from the reporters, especially the Discovery Channel has been great, and a few people from HMP camp. Yesterday Larry Lemke and Carol Stoker helped, and even Brian Glass late in the afternoon. Matt finally conceded to me that we should use the block and tackle and get some mechanical advantage in hauling up the dome segments. Before he was insisting that it could be done by hand, because they only weigh 350 pounds. As a result of using the block and tackle, things have gone a lot smoother in terms of getting the dome panels up. There is a problem: they don't seem to fit that well to the habitat. We really have to do a lot of work to force each one into place and bolt it in. My major concern right now with the completion of the habitat is that when the dome sections finally reach the edge they won't fit.

RAISING THE DOME

The procedure we had used after the walls were up was to partially build out the two decks, using wooden joists, joist hangars, plywood, and nails that Aziz scrounged for us. We then set up a small scaffold on the upper deck, and attaching a beam and pulley to it, used it to haul the dome sections right through the big gaps we had left in the floors to bring them to the upper deck. It was during one of those hauling attempts, on the twenty-fourth, that the scaffold had crashed, causing me to be hit in the head by the steel hook and block and tackle of our primitive hoisting device. On the twenty-fifth, despite my attempt to freshen the crew by quitting early the night before, we had another mistake, and the scaffold crashed again. This time Frank Schubert and Larry Lemke were both hit. Larry was not seriously hurt, but Frank's right hand took a crushing blow, which significantly impaired his ability to use tools for the remainder of the job.

Despite this second accident, on the twenty-fifth we finally succeeded in raising the arch of the roof dome. The technique used was to raise a small scaffold in the middle of the upper deck. We built a small wooden platform on top of this of just the right height, then placed the core hub piece of the dome on top of that. Then we attached a dome section panel to each side of the hub, affixing the other end of each dome section to the cylinder wall, thus forming an arch. Following this, on the same day, we added in four more dome sections, leaving six left to go.

Wednesday the twenty-sixth proved to be a banner day. We had mastered the technique for raising dome sections, and by 2 P.M. had added four more,

Pascal Lee gives briefing to HMP camp in Haughton Crater, summer 1999.

July 17, 2000. With Frank Schubert's ATV pulling and a crew of Mars Society volunteers pushing, the ad hoc "Kunzmobile" is used to haul the station's wall panels out of the crater.

July 20, 2000. “Like a dead man rising” the first set of the station’s wall panels go up.

July 21, 2000. Using a rickety scaffold, gangs of volunteers pull up the walls of the station.

July 25, 2000. The station's roof dome arch is formed.

Evening of July 26, 2000. The work team shows V signs for victory; the dome is up! Shown celebrating, from left to right: Joe Amarualik, Joannie Pudluk, John Kunz, Frank Schubert, Matt Smola, Bob Nesson, Robert Zubrin.

Nine PM, July 28, 2000. Robert Zubrin gives speech commissioning Flashline Station. Pascal Lee is standing at right.

Flashline Station, seen from the air, 3 AM, July 8, 2001. The crater lies south (above in the photo) of the snow-covered ridge slope.

Robert Zubrin, Vladimir Plester, and Katy Quinn prepare to leave station for a motorized EVA, early July 2001. An equivalent to the ATV's used in the Arctic will be a necessity for Mars explorers.

July 16, 2001. Members of Flashline Crew 2 struggle to free a trailer carrying geophone seismic equipment from thick mud in Haughton crater. There will be no mud on Mars, but there could be sand traps. The hard physical work necessary to deal with such situations underscores the need to employ artificial gravity on Mars-bound spacecraft.

After the incident with the mud trap, the EVA team debriefs for after-action analysis. Shown are geophone principal investigator Vladimir Plester, NASA Ames human factors researcher Bill Clancey, and crew commander Robert Zubrin.

July 20, 2001. On the 25th anniversary of the Viking landing, geologist Charles Frankel encounters the dormant Carnegie-Mellon Hyperion rover on the Von Braun Planitia north of the station. Perhaps in the not-too-distant future, an astronaut will have a similar meeting with one of the Viking landers now resting on Mars.

Austrian geologist Katy Quinn wields a sledge hammer to initiate subsurface signals for geophone analysis of Haynes Ridge. Stratographic data was obtained down to 1600 ft (500 m) below the surface.

Charles Frankel and Brent Bos climb Marine Rock, a large outcrop standing astride the Von Braun Planitia. The human explorers' ability to climb not only provides access to the geology of the hill top, but gives the explorers a useful view of the surrounding territory.

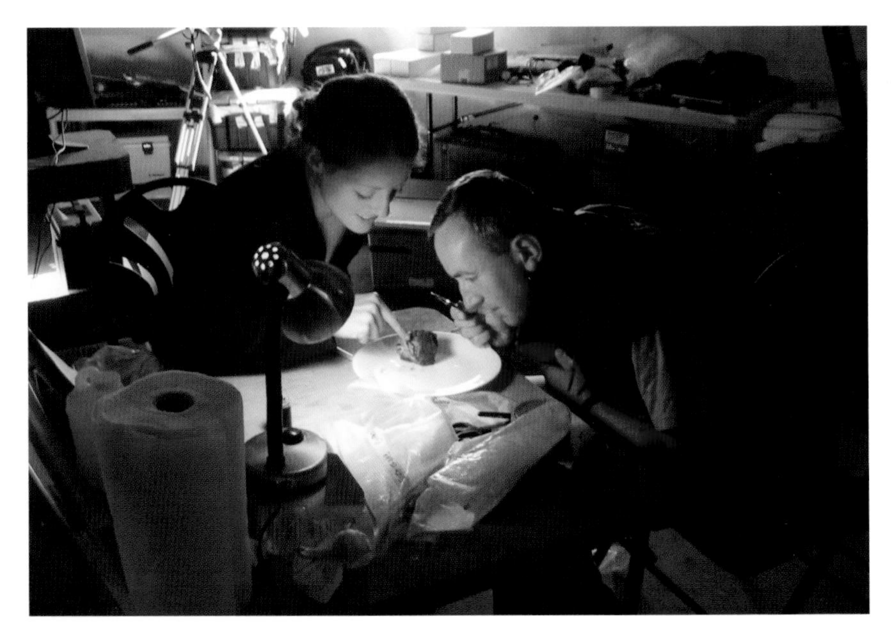

Charles Frankel and Danish planetary scientist Cathrine Frandsen examine rock samples in the Flashline Station lab during Flashline Station Crew rotation 3.

The DARPA-US Army experimental reconnaissance telerobot "Solon" goes over the top to explore Devo Rock canyon on Devon Island, July 26, 2001. Solon survived his descent into the canyon, but was destroyed six weeks later during rescue operations at the World Trade Center.

John Blitch, Cathrine Frandsen, and Brent Bos command Solon from a mobile field station at the rim of Devo Rock Canyon, July 26, 2001. Small telerobots will be more useful to human explorers than large independent robots, because they can be flexibly deployed by highly mobile field teams and sent where humans cannot go. *(courtesy Discovery Channel/Resolute Films)*

February 2002. A motorized EVA team leaves the Mars Desert research Station during MDRS crew rotation 1.

Heather Chluda and Steve McDaniel scale a steep hill during MDRS crew rotation 1, February 2002.

Jennifer Heldmann, geologist for MDRS crew rotation 1, explores a rocky hillside. Scientists working on Mars will need spacesuits that allow for the kind of mobility required to explore such difficult terrain.

Robert Zubrin and Heather Chulda survey sheer cliff formations several miles north of the Mars Desert Research Station. Studying cliff faces is useful, as they reveal stratigraphy. Triathlon athlete Heather expressed a desire to rappel down the cliff, but was overruled. The job was later done by a cliff-scaling telerobot developed by the French Mars Society.

July 2002. Members of Flashline Station Crew 7 exit the station to begin EVA.

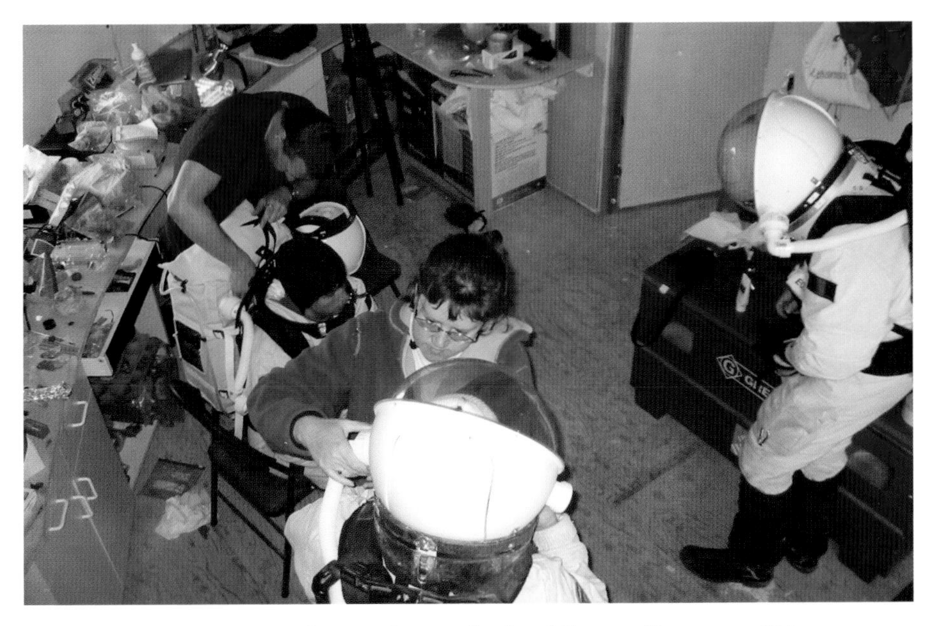

Flashline Station Crew 7 members Markus Landgraf and Shannon Hinsa assist EVA team members in suiting up. While doffing suits could generally be accomplished in the stations EVA preparation room, the more complex operations required to don the suit typically required the use of the station's lab area as well. In order to limit dust entry into the station, suits would need to be cleaned in the EVA preparation area prior to being brought into the lab.

Flashline Crew 7 geologist Nell Beedle examines fossilized algal mats found on the side of rocks at the bottom of Devo Rock canyon.

July 2002. After assessing its value as a potential radio repeater site, members of Flashline Crew 7 descend from the summit of Lakeview Hill.

Markus Landgraf uses a geologist hammer to obtain rock samples for analysis by the Carnegie Institute MASSE instrument.

July 2002. Members of Flashline Crew 7 explore Devon's polar desert on ATVs. The featureless nature of desert landscapes put a premium on electronic navigation systems as a guide to explorers.

Flashline 7 crewmembers Nell Beedle, Emily MacDonald, and Frank Eckardt use a reflectance spectrometer to provide the farthest north ground truth measurements ever taken for the JPL MISR instrument currently orbiting the Earth and NASA's Terra Earth observing satellite. By combining the abilities of human surface explorers with orbiting satellites, the scientific return of both can be multiplied.

leaving only two to go. The problem, however, was that unlike the previous dome sections, which had all had at least one open edge available at the time they were raised, these last two would have to be raised and placed into pie sections bordered on each side by previously raised dome panels. This made things much tougher. We started our attempt to raise the eleventh dome section around 2 P.M., and it was only after repeated failed tries and five hours of shoving and pushing and hammering that we finally got it in place and bolted down.

Around 6:30 P.M., as the effort on the eleventh panel neared conclusion, the Inuit boys started to whistle, in a very weird undulating kind of way. I asked them why they were doing this, but they just smiled and kept whistling.

It did not, however, take long for their meaning to become clear. A few minutes after the boys started to whistle, the wind began to rise, the sky became overcast, and a light rain began to fall. Our eight-day lucky streak of fine weather was at an end. Somehow, the Inuit had seen the signs of the weather change in advance.

That brought us to 7 P.M., with one dome panel left to go. The team was dog-tired, and if the eleventh panel was anything to go by, the twelfth would be no picnic either. But with the weather changing, there was the threat that a high wind could get in under the partially completed dome and exert massive force to rip it off the hab. We decided to make a go for putting in the final dome section.

Frank and Matt, our strongest and most skillful workers, got up on the roof to try to guide the dome section from there. The rest of us hauled the dome panel to the top of a small scaffold and then lifted it from below to push it through the last remaining open pie section in the roof. Frank and Matt grabbed it from above and steered it into what they hoped would be the correct direction. Then, at the signal, we rammed the 350-pound panel down and forward. Miraculously, we were dead on target and the thing slid right into place like a hand into a glove. It was tremendous. "It's in!" we cheered, and rushed to slam in bolts to make sure it stayed there.

Of course, that left Frank and Matt marooned on top of the now completed dome. But they had a rope and rappelled down to climb into the upper-deck porthole window.

It was done. The dome was up!

THE FIRST SALUTE

With the completion of the dome, the dangerous part of the work of building the station was over, and the number of volunteers who turned out to help increased dramatically. In fact, except for Pascal, a couple of other hard-core HMP honchos, and a reporter from the macho magazine *Men's Journal,* virtually everyone from camp came out to the hab and pitched in. Before, our camp volunteers had consisted mostly of dedicated Mars Society members like Larry Lemke, Carol Stoker, Bill Clancey, and Darlene Lim, plus off-duty members of the Discovery Channel film team. Now we had half a dozen unaffiliated NASA scientists, an equal number of university people, industry engineers from the Hamilton Sunstrand Company (who turned out to be great carpenters), and journalists from a pack of media outfits. The weather outside had turned bad, but the hab's inside was now sheltered. As we built out and painted the interior and finished building out the decks, the atmosphere was akin to a barn raising. Progress was rapid, and it became apparent that we would be able to commission the station soon.

On the evening of July 27, I sent an e-mail message to the Mars Society Mission Control in Denver to establish contact in preparation for the commencement of simulation operations the next day. "Mission Control, this is Flashline Station. Are you there? Please Respond."

Mission Control replied, "Flashline Station, this is Mission Control. It's good to hear from you. Clearly, failure was not an option."

That night, Frank and I slept in the hab for the first time. We did not inform Pascal.

While straightforward in principle, the amount of interior work that needed to be done exceeded the time available. By the evening of the twenty-eighth, the hab interior was still not finished. In fact, it was a complete mess. The bunks were built out and painted white, but except for splatter from the bunk jobs, the floors weren't painted at all, and not all of the interior partitions were installed. Nevertheless, since Frank, Matt, and I were leaving the next day, we decided to go ahead with our plan to inaugurate it and have our first symbolic occupation that night. Further improvement work could be done during the stay of the short simulation crew led by Carol Stoker that would follow us.

So an inauguration ceremony was scheduled for 9 P.M. at the hab, and the entire population of the island, consisting of some fifty scientists, Inuit, and journalists, turned out to attend. After several others spoke, I gave the concluding remarks. I dedicated the station to those whose cause it will ultimately serve, a people who are yet to be, the pioneers of Mars.

Then, at my signal, John Schutt fired a shotgun into the air, as a first salute to *their* flag, the red, green, and blue Martian tricolor that we had raised that day to fly atop the station. At the sound of the blast, a cheer went up from the crowd. Something grand had been born.

I was then given a bottle of champagne, which I smashed against the habitat to christen the station. This provoked a sigh from the crowd, who had spent the summer pretty dry, and who no doubt had other thoughts as to the potential uses of the fluid I had just splattered all over the habitat steps.

Pascal, however, immediately reassured them: "It's all right, folks. It's just Canadian champagne."

Then the symbolic first crew, consisting of Pascal, Marc Boucher, Frank, HMP biologist Charles Cockell, the Discovery Channel's Bob Nesson, and I, entered the hab. We sat down at the table, had a meal, and fired off a few e-mail messages to Mission Control via the Internet link that the HMP's communications director, Steve Braham, had run into the place. Then, as Frank got out his guitar to strum a few chords of a Beatles song, I looked out the window.

It was a bittersweet moment for me. Against the odds, we had been successful. The hab was up. While its interior was not complete, it would be fitted out in time. Yet to the extent there was any simulation operation that year, I would not be part of it. I also had another feeling that must seem even odder to most readers of these lines: I was wistful that the construction battle was over. An adventure had come to an end.

The next morning, we had time for a brief simulated traverse. We had no spacesuit simulators yet, so we adopted the simulation rule that a spacesuit would consist of an ATV helmet, a backpack, a parka, and thick ski gloves. While you were outside, it was forbidden to remove any of these elements, and you could only talk to others via handheld FRS radio. It was surprising how much change such a simple "suit" created in the relationship between its wearer and the environment. For example, it suddenly becomes a matter of some difficulty to manipulate the tiny buttons on a camera when you cannot take your ski glove off.

The weather had turned sunny again, and we set off on an ATV traverse led by Pascal. He took us to a ledge overlooking a spectacular canyon system with a huge rock at its bottom. Frank promptly christened the boulder Devo Rock, after his former band.

I saw two seagulls gliding in the canyon, the first birds I had seen since my arrival on Devon Island. The soaring flight of seagulls has always been for me kind of a vision of joy and freedom, and they seemed to symbolize the exhilaration of that magnificent morning. After two solid weeks of brutal and dangerous work, we were roaming the countryside on ATVs, as free as birds.

It was over too soon. A short while after we left the canyon, we heard a droning sound in the sky. It was a Twin Otter, flying in to pick up Frank, Matt, and me for our return flight to Resolute Bay and points south.

First Air Twin Otters wait for no man. We gunned our ATVs and dashed back to camp to pick up our gear in time to catch the plane.

Carol Stoker had heard the Twin Otter too, and knowing the problem we faced, had packed up our tents for us and brought our gear to the airstrip. In my tent, she had discovered my summer reading. "I know your secret," she said as she handed me the volume.

It was a book about ancient Greek and Roman engineering.

When we reached Resolute Bay, Aziz was waiting for us, and he took us to his hotel for a very badly needed shower. Greg Clark, a reporter for Space.com who had come to the Arctic to cover our effort but had gotten there late, saw us and interviewed us. Then it was back to the airport for the 737 flight to Yellowknife, the capital of the Northwest Territories.

At Yellowknife there was a one-hour layover. Matt and I walked around the airport and then reboarded the plane, but Frank was nowhere to be found. When the plane took off, it left him behind. We found out later what had happened.

Frank had found the saloon.

7. PREPARING THE SIMULATION

There were still five days left in the HMP field season after Frank and I departed, sufficient time to attempt a short shakedown mission simulation. So we did one.

The shakedown crew was commanded by planetary exploration engineer Carol Stoker, of NASA Ames Research Center, and included aerospace engineer Larry Lemke and human-factors scientist Bill Clancey, both also of NASA Ames; University of Toronto paleogeobiologist Darlene Lim; Mars Society Canada President and Internet entrepreneur Marc Boucher; and the Discovery Channel's Bob Nesson. This shakedown crew lived and worked in the hab for four days, supporting a series of exploration traverses on Devon Island. The spacesuit simulators employed were limited by available resources to the same helmet-backpack-gloves-radio convention the symbolic first-night crew had adopted during our brief sojourn. However, the Hamilton Sunstrand corporation (which makes NASA's spacesuits) had brought to the Arctic a sophisticated prototype of an actual Mars spacesuit, and this was made available to the crew for field testing. To report of their activities, the crew engaged in dialogue with Mission Control in Denver, with twenty-minute time delays inserted before response to duplicate what would occur in radio communication between Earth and Mars.

The hab was barely functional at this time. The primary structure was up, and while they did not look pretty, the walls and interior partitions for the personal staterooms and the airlock were all in place as well. There was electricity for lights, hot plates, and recharging devices, provided by a gasoline generator parked about 20 meters away, but no running water. A camp toilet (crap in a bag) was used to deal with metabolic waste. There was no refrigerator, so food was shelf-storable dried items plus military MREs (Meals Ready to Eat). A com line had been installed linking to the satellite dish of the base camp, enabling e-mail correspondence. But use of this connection was impaired by whimsical decisions on the part of someone in the HMP to limit crew access to this link to four hours, then two hours, and finally one hour per day.

Given these limitations, the shakedown sim was crude in the extreme. Nevertheless, observations of a group of planetary scientists and engineers attempting field exploration and reportage under these conditions provided good data for Bill Clancey's human-factors studies. The simple experience of living in the hab for four days enabled identification of a long list of items needed for correction, installation, or improvements. The work with the Hamilton Sunstrand spacesuit was a good example of the kind of exploration technology testing the station could undertake in the future. Observations were also made by the Mission Support Team (as the Denver-based group formerly known as Mission Control appropriately renamed themselves), greatly reinforcing previous Rocky Mountain Mars Society data supporting the need for well-organized mission communication protocols.

Carol and company left the island on August 4, ending sim operations for the year. A few days later, the Mars Society held its Third International Convention in Toronto.

The atmosphere at the convention was triumphant. Everyone was aglow with our miraculous comeback to success on Devon Island. At the Steering Committee meeting Saturday afternoon, I pushed through a resolution that we seize the time and capitalize on our momentum by expanding the Mars analog research station program to four stations. In addition to the one on Devon Island, we needed others in the American desert, Iceland, and the Australian outback.

The American desert station offered the advantage of maximum ease of access and available yearly duty time, thereby enabling a great expansion in both the quantity and variety of work that could be accomplished on Devon alone. Since it would be in the northern hemisphere, it could operate fall, winter, and spring—exactly the times when Flashline Station was inoperable, thereby affording us the capability for a year-round program of Mars analog field research.

Iceland offered an excellent Mars analog natural environment of a geologic type different from Devon, being volcanic and hydrothermal instead of impact/sedimentary. Moreover, it was in Europe. We needed a station in Europe to wake up the continent of science's past to science's future. Europe collectively has a population and economy equivalent to that of the United States, yet its contribution to space exploration has been pitiful even in comparison with NASA's rather mixed-quality effort. That needed to be changed, and the European Mars Arctic Research Station or EuroMARS, could begin to do that.

Finally, the Australian desert is known for its findings of the oldest fossils on Earth. Dating back 3.5 billion years, the Australian stromatolites (fossils left by colonies of bacteria) are the earliest evidence we have for life on this planet. At the time that these stromatolites were forming, there were oceans and rivers on Mars. They are thus exemplary of the kind of fossils considered most likely to be findable on the Martian surface. Searching for ancient stromatolites under simulation conditions in Australia would thus be an exact analog for similar research on Mars.

The plan was to launch one new station per year, starting with the

American desert, then Iceland, then Australia. When the program was announced at the Mars Society banquet that evening, $78,000 was raised on the spot to get it under way.

The Mars Desert Research Station (MDRS) project was begun.

There were also some developments among the people involved in the program at this time. For most of the people in the Mars Society, the successful building of Flashline Station had been a triumph. Marc Boucher, however, had opposed proceeding after the failed paradrop. While he had actually turned out to work for an hour or so a few times during the construction effort, this apparently was not enough to give him a sense of pride in the achievement. He became a malcontent and, together with his new business partner, spread much negativity. A year later, after failing to induce the international chapters to disaffiliate from the Mars Society, he finally dropped out.

Pascal's behavior in the Arctic that summer came under review by NASA. He had offended many people at camp with his overbearing manner, and there had been other problems as well.

As the NASA technical oversight monitor of the HMP's contract, Chris McKay called me up to get my take on the situation. It was true, I said, that Pascal's arrogance in the field during the summer of 2000 was almost surreal. His inconsistant support for our project had also been irritating, to say the least. Yet at a time when the other principal figures on the scene had all wanted to cut and run, he had at least held in place, and he deserved some credit for that.

The HMP had sustained a severe injury from an ATV accident. Did I think Pascal's leadership was in any way at fault? No, I did not. Pascal was a stickler for ATV safety. Overall, I gave him a mixed review. He did not need to be dismissed, I said, but he did need to be confronted and compelled to change.

Chris concurred, and Pascal remained in place.

I had cause to doubt the wisdom of my advice a few weeks later, however, when Pascal called me up to talk about the Marine Corps paradrop bill. There had been four Herc flights, and, Pascal said, the Marines were asking $25,000 for each, or $100,000 total. According to our 90/10 split agreement, that meant we should pay $90,000 and the HMP would pay $10,000. He requested that I wire the $90,000 into a personal bank account that he shared with one other individual, and he would then use it to pay the Marines. I told him we were being sued by the crane company for the loss of their equipment, and the cost of the settlement, as yet undetermined, would need to be deducted, along with the costs of the other equipment lost from any transportation bill. He said, very well, in that case send $45,000 now, and we could work out what the required balance would be on the rest later.

I agreed to this, but instead of wiring the money as requested, I mailed him a check for $45,000 made out to the United States Marine Corps. As soon as he got it, Pascal called me up at home and demanded that I replace the check with one made out to him. This I refused to do, and he became quite agitated, saying that the Marines could not accept a check from the Mars Society, so the check I sent him was useless and absolutely had to be replaced. I would not budge, and terminated the conversation.

The check, made out in early September, did not clear our bank account for four weeks. When I finally got it back from the bank, bank stamps imprinted on it showed that the Marines had cashed it on October 4, but it appeared that, prior to that, the check had been unsuccessfully deposited without endorsement. This was evidenced by the front of the check, which was stamped "Reason for Non Payment: Endorsement not as shown."

These facts led me to suspect that Pascal may have tried to deposit the check in his own account. It is possible he may have wanted to do this just so he could pay the Marines himself. Nevertheless, this made me very uncomfortable.

SCOUTING THE GREAT AMERICAN DESERT FOR MARS

The first task in the development of the Mars Desert Research Station was to find a site. We wanted a place featuring a large uninhabited and unvegetated area within which crews could operate without excessive human intrusion. The region needed to have varied geology representative of at least some of the terrain types of Mars, making it interesting to explore and a useful test bed for Mars mobility systems. We also wanted it to be relatively free of snowfall, so that the geology would remain exposed during winter, yet not so hot as to preclude operations in simulated spacesuits for the majority of the year. Finally, or perhaps firstly, it had to be in a place where permission could be obtained to establish a base.

The American West is still mostly uninhabited, but it is all owned by someone. There are several major categories of land, including National Parks, National Forests, Federal Wilderness Areas, Military Reservations, Department of Energy Reservations, Indian Reservations, Bureau of Land Management (BLM) lands, state lands, and private lands. The issue of getting permits absolutely ruled out National Parks, and argued strongly against National Forests and Federal Wilderness Areas, but all the other categories were potentially viable. The amount of territory to explore was huge, roughly a million square miles. The only way to do it was to mobilize the membership of the Mars Society.

We constituted a Southwest Task Force, and sent groups of Mars Society members on scouting trips across large areas of west Texas, New Mexico, Arizona, California, Nevada, Utah, Colorado, Idaho, and eastern Washington. One of the earliest expeditions was led by Chris McKay and included me, my wife Maggie, our daughter Rachel, and Frank Schubert. We went to the Mojave Desert, an area that Chris had explored extensively as part of his search for endolithic bacteria. The area was scientifically interesting, but it was still too hot in October, signifying that its useful working season for crews in simulated space suits might be far too short. Moreover, as the presence of weekend dirt-bike enthusiasts showed, it was too close to Los Angeles to provide sufficient privacy.

Jon Wiley scouted west Texas, but the areas he searched were too vegetated. Shawn Plunkett, a Mars Society member who is a pilot, scouted Lunar Crater Nevada from the air, and it looked good, but a subsequent ground-based examination showed that what seemed unvegetated from the air was manifestly covered with scrub brush when seen from the ground. Stacy Sklar found some promising sites in the Painted Desert of northern Arizona, but these were on Navajo land. When I called up the Navajo tribe to find out their terms for access, a spokesman explained that they had a fourteen-step approval process for any project, and a single member of the tribe could veto the project at any step. That ruled out the Painted Desert. The White Sands Missile Range development people heard about our project, and invited us down, but when our members checked it out, they found that most of it was too vegetated for our purposes. We were also invited to check out Area 51 in Nevada, the alleged center of U.S. government flying saucer research, according to UFO enthusiasts. This site would certainly have added a bit of pop culture panache to our project, but alas, the fabled area turned out to be too Earthlike for our tastes. I did obtain a flying saucer driver's license from a nearby souvenir shop, however. This frequently comes in handy when I am asked to show identification at airport security checkpoints.

In an effort to develop some kind of rapport with Pascal, I invited him to join me for a two-person scouting trip through southern Utah in November 2000. Pascal is a good geologist, and I found his explanations of the various exotic landforms we encountered quite interesting. The effort at bridge building failed, however, when we came around to discussing our philosophies of life.

I believe that humans have free will, and that by using it we can define a purpose for our lives whose validity is demonstrated by the good we leave behind. I also believe that humanity can, if it chooses to, have a purpose, whose validity will be demonstrable by the good we create in the cosmic sense. Thus we need to explore the universe in order to discover what our purpose might be, and act to advance the powers of human civilization so as to make us capable of implementing it.

Pascal, on the other hand, denied the existence of free will. According to his explanation during our trip, what you believe are your own decisions are actually your predetermined responses to external stimuli. Therefore, there can be no such thing as purpose, good or evil. These ideas are all metaphysical constructs. Things happen, and humans, like all other physical phenomena, respond mechanically in accord with the laws of science. We do not make any decisions, we simply observe what decisions we are made to make. I found this view of the world incomprehensible.

We did, however, have an interesting experience. Arriving in the town of Page, Arizona, in the evening after a long drive, we encountered huge mobs of panicked people being herded down the main street by U.S. Army troops. This caused us some concern, until we noticed that the TV crews filming the disaster were not adorned with the insignia of any news organization. In fact, they were film cameramen; we had wandered onto the scene of a movie set. We discovered later that the film in question was the David Duchovny alien invasion farce *Evolution.* If you see this movie and observe two men pushing sideways through the fleeing crowd to try to reach the restaurant across the street, that's us.

Another unusual scouting expedition was actually sponsored for us by the BBC. They were doing a documentary about Mars exploration, and wanted to feature the Mars Society. Hearing that we were scouting the American Southwest for the site of our next station, they thought it would be cool to get footage of us doing this—from balloons. They offered to fly me down to Monument Valley in northern Arizona if I would do the honors. The idea was absurdly Jules Verne-ish, but it sounded like fun, so I accepted. We took off in a hot-air balloon and promptly bumped right into the side of one of the big monument rocks. Then, gaining altitude, we climbed up and over the giant rock, to soar grandly above the spectacular valley. That part was exhilarating. When it came time to land, however, we still had a lot of ground speed, and all of us were tumbled right out of the basket when the gondola touched down. The producer was scared out of his wits, but everyone else had a good time.

In the end, the tip that led to the discovery of the actual site for the Mars Desert Research Station came from an unusual source. James Cameron, the filmmaker who produced the movie *Titanic,* is a Mars Society member. In 1999, he spoke at our Second International Convention in Boulder and wowed the crowd with the graphics of the exploration vehicles he had conceived for a 3-D Imax Mars movie he was planning. As part of this movie project, Cameron had sent his own scouts out to scour the country to try to find a place that most resembled Mars to use for a set. When he heard we were looking for Mars in the American Southwest, he contacted us and suggested two places. One was the Painted Desert. We had already found that one ourselves but lost interest after encountering the Navajo approval process. The other site was the desert area west of Hanksville, Utah. Cameron and his people were even more enthusiastic about this area. "It's Mars!" Mike Novotny, his search coordinator said.

We sent our intrepid Southwest task force member Jon Wiley to the area to check it out, and he found some places that seemed somewhat interesting. Another search team from Colorado then investigated the area, but through a navigational error missed the sites that Jon had found, and came back with a negative report. In order to resolve the discrepancy, Frank Schubert and I went out. We found the place that Jon had cited, and in principle it did offer a lot, but Frank judged it too hard to build on. By this time, I had already talked with both the federal BLM and the Utah State Trust Lands Administration people in charge of the area, and had found the state authorities, represented by one Lou Brown, much more receptive to hosting our project. We therefore continued the search, focusing on the scattered squares on the map that corresponded to Utah State land.

Eventually, we found two good sites. One was in an area of gray Cretaceous marine sediments not far from the spectacular rock formation known as Factory Butte. This was in the same geological unit that Wiley had identified. The other was in an area of red Jurassic terrestrial sediments a few miles farther to the east. Both areas were fossiliferous, making them interesting zones for simulated exploration missions, and both featured varied terrain that would be useful for testing robots and human mobility systems. The area was located 4,500 feet above sea level, in a climate cool enough to permit operations fall, winter, and spring, exactly as we had hoped, and the snowfall was low. The population density of the surrounding area was minimal, affording the privacy we would need, but the friendly attitude of the folks in Hanksville assured us we would be able to find help if we should need it as well.

It was tough choosing between the gray Cretaceous and the red Jurassic sites, because both had a lot to offer. Ultimately logistical considerations influenced me to choose the red site, as the better dirt road leading to it would make it easier to resupply by truck. By spring 2001, the choice had been made.

The MDRS would be located in the red desert a few miles northwest of Hanksville.

THE CALL FOR VOLUNTEERS

In the fall of 2000, the Mars Society Steering Committee began to plan the mission simulations to occur on Devon Island the following summer. At the top of the list of issues was the selection of the crews.

It was clear that we had available to us in the established Mars research network sufficient numbers of people to fully staff Flashline Station for the coming field season. But I believed very strongly that there was a much larger group of highly talented individuals from diverse fields who were currently unknown to us but who had a great deal to offer our program. If presented with a chance, I said, these people would line up in droves to volunteer and add their skills and their passion to our efforts. It would be foolish to turn our back on this potential and keep the Flashline

Station program as an in-crowd affair. There was some dissension from some Steering Committee members on this point, but Chris McKay backed me up, and so ultimately my proposal for an open field program prevailed.

Accordingly, in November 2000, the following advertisement was sent out over the Internet.

VOLUNTEERS NEEDED FOR FLASHLINE CREW:
HARD WORK, NO PAY, ETERNAL GLORY

The Mars Society is requesting volunteers to participate as members of the crew of the Flashline Mars Arctic Research Station during an extended simulation of human Mars exploration operations on Devon Island during the summer of 2001. It is anticipated that the field season will run from late June through late August; volunteers should state what segments of this span they are available. Both volunteer investigators who bring with them a proposed program of research of their own compatible with the objectives of Flashline Station and those simply wishing to participate as members of the crew supporting the investigations of others will be considered. Applications will be considered from anyone in good physical condition between 18 and 60 years of age without regard to race, creed, color, gender, or nation. Scientific, engineering, practical mechanical, wilderness, and literary skills are all considered a plus. Dedication to the cause of human Mars exploration is an absolute must, as conditions are likely to be tough and the job will be very trying. Those selected will be required to participate in certain crew training exercises to take place in the western United States during the spring of 2001, and to act under crew discipline and strict mission protocols during the simulation on Devon Island. There will be no salary. Applications including résumé, character references, and a brief letter explaining why you wish to participate should be sent to Mars Society, PO Box 273, Indian Hills, CO 80454 no later than Jan 31, 2001. Total length of applications should not exceed 3 pages. Please include 7 copies.

The response was astonishing. Over 400 applications arrived. The applicants came from all over the world, and every walk of life. We had engineers and field scientists, but also military officers, adventurers, philosophers, artists, musicians, reporters, businessmen, anthropologists, economists, medical doctors, lawyers, students, clergy, and members of the congressional staff. We had the soldier of the year of the Israeli Army and a special agent for the Hague War Crimes Tribunal who had conducted behind-the-lines operations in Serbia.

Choosing crew from this incredible array of talent proved difficult. Out of the 400-plus applicants, more than 300 were viable choices as potential crew members, and dozens were so good that it was positively painful to turn them away.

We desired cooperation from the Haughton Mars Project, and this would only be forthcoming if Pascal were given command of the station for a large part of the summer. Since at this time, Pascal represented himself as a member of the Mars Society who would lead mission simulations on our behalf, this did not seem that much of a sacrifice. So command of the station for the 2001 field season was divided between Pascal and me, with Pascal getting twenty-six days, which would encompass rotations 1, 4, 5, and 6, while I was allotted twenty-one days to lead the somewhat longer rotations 2 and 3.

There were thus six crew rotations with five open slots each, creating thirty potential openings in all. HMP radio system overseer Steve Braham wanted to be in the hab for all six rotations, however, as he wished to place his communication gear there. Since we needed a com system, I agreed to this on condition that he sign a contract guaranteeing the station full access to his communication link for the entire summer. He did, leaving twenty-four slots open. Pascal was strongly biased toward choosing HMP members for crew assignments, while I wanted to maximize the opportunities for newcomers. After some rather tedious negotiations, a compromise was reached, with about half the open slots going to the HMP crowd and half allotted to the volunteers. The Flashline Management Committee then conducted a series of downselect votes to reduce the pool of volunteers to about sixty and then to twenty, after which Pascal and I made our selections from the final short list to fill out our crews. At the end of it all, between the HMP and the volunteers, twenty-five people were chosen for crew slots. The twenty-five included nineteen men and six women, with the United States, Canada, Australia, Britain, France, Belgium, and Denmark all being represented.

LEARNING TO COMMUNICATE

The selection process took all winter and was not completed until March. While this was going on, the Rocky Mountain Mars Society (RMMS) was doing its part to prepare the simulations by continuing its own communication protocol development missions into the mountains and engaging in an extended effort to design, build, and field-test spacesuit simulators for the crews to use in the Arctic.

The problem identified by the RMMS's Devon Island 2000 Mission Support experience was that it is extremely difficult for a rear-echelon group that has never been to a place that a field team is exploring to gain an adequate understanding of what is really going on. It was not merely a question of capricious HMP constrictions of the link time. That was admittedly a contributing factor, since had the link been kept open longer, Mission Support could have interrogated Carol's shakedown crew with a series of interactive messages to draw out more data. The real problem was that the crew did not know what to report. They knew what *they* were experiencing, but they could not see what their mission looked like to the people on the Denver end of the com link. There are different dimensions of the crew's experience—such as the narrative of the day's chronology, the feel of the occurrences, the scientific observations, and the ongoing wrestling match with the always cranky engineering systems—that *all* need to be communicated if Mission Support was to have any understanding of the actual situation.

There is thus a minimum set of reports of specific different types needed to adequately convey the day. While the folks back home would like the reportage to greatly exceed this minimum, a requirement for excessive reportage would impair the actual operations of the crew. Based on its experience with the shakedown crew, the Rocky Mountain Mars Society was able to greatly improve the focus of its program of communication protocol design.

After a series of test missions into the mountains conducted over the winter of 2000–2001, the RMMS Mission Support group came up with a daily requirement for four categories of reports.

First, Mission Support wanted a summary report on the mission's overall operational status, tersely describing what the crew did that day, what they planned to do the next day, and what problems they faced. As this tended to usually come from the Commander, this report became known as the Commander's Report.

Second, Mission Support wanted a Science Report, containing the scientific observations made in the field and in the station's lab on a daily basis. This recommendation was to encounter opposition from some of the scientists included in Flashline and later MDRS crews, as daily observation reports are not ordinarily filed in the course of scientific field or laboratory work on Earth. Rather, the way terrestrial scientists are accustomed to working is to go into the field and spend a month studying a region, returning to their university lab to examine their collected samples and photos for six months or a year. Only then, after having subjected their data to such mature consideration, would they release their results in the form of a paper submitted for publication in a journal or presentation at a conference.

As standard as such procedure might be for fieldwork on Earth, however, it would never pass muster on the Red Planet. There is simply no way NASA would send a team of explorers to Mars and wait to see what they found in a paper published a year after their return. They would insist upon receiving daily reports so they could acquire the data as it unfolded. The scientists might deride this requirement as an expression of the philistine proclivities of uneducated bureaucrats who know nothing about how true intellectuals need to work, but in fact, the managers would be right. Near real-time transmission of scientific observations is a necessity, because unlike the field scientist on Earth, the scientist on Mars is not operating as an individual; he or she is the point of a spear. No one person can be a qualified specialist in all the areas required to do a field scientific investigation of Mars. Therefore rather than acting as a lone principal investigator of the classic type, the scientist on Mars must act as the frontline member of a team of scientists incorporating all the disciplines of the terrestrial scientific community. That is not to say that the Mars scientist will have his or her activities dictated by an Earth-based committee. There is a reason why we prefer to send humans to Mars instead of robots, and if we are to take advantage of the full potential of a human investigator on the scene, we don't want to enslave her to NASA's Mars Science Working Group. Quite the opposite, we want to enslave the Mars SWG to her, or rather, since we also want them to use their minds, enlist them in her research company. But in order for them to serve her properly, they need to see the data that she is seeing, so they can apply their array of specialized knowledge to it, and support her work in real time. The investigation needs to be led from the field, but given the cost of a human Mars mission, the Mars-based generalist acting as overall Principal Investigator should have a team of Earthbound specialists serving as her staff. This puts a reportage burden on the scientist that she may not be accustomed to, but accepting that burden will greatly enhance her capability overall. It allows her to be supported through *telescience*.

Depending on the character of the work being done, there might be one science report, or several, with each devoted to separate areas, such as geology, biology, and paleontology.

The third important report needed by Mission Support is the Engineering Report, covering the status of systems and what efforts are being made to maintain or repair them. In an actual Mars mission, the Mission Support group would unquestionably have at its disposal data readings from automated engineer health monitoring systems, but human discussion of the behavior of the engineering systems in the field ("the wheels were knocked severely out of alignment but it still can be driven") and the proposals for repair attempts will still be necessary. Useful advice from Mission Support on how to effect necessary repairs in equipment is generally appreciated by the field crew, and so we usually don't see much resistance to the filing of Engineering Reports.

The fourth and final report defined as necessary by Mission Support was the least conventional but perhaps the most important. This was the Narrative Report, or as I and some others prefer to call it, the Journalist's Report. By "journalist," I do not mean the term in the twentieth-century sense of a member of the commercial news media. Rather my journalist is a journalist in the eighteenth-century sense—he or she is a crew member who keeps a diary. People today can still experience the expedition of Lewis and Clark, because Lewis kept an excellent journal. Many other famous expeditions and voyages of exploration have been basically lost to human experience, because no proper journals were kept (or preserved). Given the importance of the human mission to Mars, it is essential that a person capable of putting the sensuous daily experience of the crew into words be included. The journal this man or woman of letters writes will not merely be of historical significance, however. It also will supply Mission Support with its most holistic view as to what is actually going on in the mission.

Another important aspect of communication over time-delayed transmissions is the need to assure each side that its messages are actually being received on the other side. Mission Support's recommendation for handling this was for each communiqué to be numbered in sequence, and for each to be acknowledged with an "Ack" message upon receipt.

SIMULATING SPACESUITS

The development of the spacesuit simulators was a project that involved close to a year's work by a group of RMMS volunteers led by Dewey Anderson and including Lorraine Bell, Tony Muscatello, Robert Pohl, Brian

Enke, Patty Piteau, and me. We needed a suit for crew members to wear that would duplicate to a substantial extent the loss of mobility, agility, dexterity, visual capacity, and situational awareness an astronaut explorer wearing a real spacesuit might experience on Mars. We also wanted it to be of sufficient complexity that the time and care required to don and doff it would also mimic to some degree those required by a real spacesuit, and thus constrain the scheduling of crew activities in a similar fashion. For purposes of both crew psychology and public outreach, we wanted our simulators to look like real spacesuits. Finally, and most critically, the spacesuits had to be safe and functional for use in the actual Arctic environment where they would be deployed.

Strictly speaking, it is impossible to build a simulator that fully duplicates the user impact of a Mars spacesuit on Earth. This is so regardless of the program budget available or the assumptions made about the actual Mars spacesuit, because the gravity of Mars is irreparably different from that of the Earth. For example, a true Mars suit might have a mass of 75 kilograms (165 pounds) and the astronaut using it might also have a mass of 75 kilograms. On Earth, most people would have difficulty walking anywhere wearing a 75-kilogram suit. But Mars's gravity is 38 percent that of Earth, so when worn on the Red Planet, the suit would only have a felt weight equivalent to 28 kilograms (61.5 pounds) on Earth. So perhaps we should make our simulator weigh 28 kilograms—this would at least be manageable by a typical physically fit astronaut. But this would still be wrong, because the 75-kilogram astronaut would also weigh 28 kilograms on Mars, and so the total load his legs would need to lift would actually be 28 + 28 = 56 kilograms (123 pounds), which is less than their burden when he is walking about stark naked on Earth. To accurately simulate its feel on Mars, the Earth-based simulator would thus need to have negative weight! Well, we clearly can't do that, so how about a spacesuit simulator with a weight close to zero? That's not quite right either: while not overburdening the lifting strength of a Mars astronaut, the spacesuit's mass would still throw off his center of gravity and moments of inertia in ways that would certainly make him much clumsier and impair his mobility that way. So there is no right solution. Our answer was to create a suit that weighs about 15 kilograms (33 pounds), which limits the mobility and agility of the user not in exactly the same way as a suit would on Mars but in comparable ways.

One area where we could, in principle, have made the suits more realistic than we did would have been by pressurizing them. Creating pressurizable suits, however, would have been *very* expensive, so much so that this step was beyond the means of the Independent Research and Development budget of the Fortune 500 Hamilton Sundstrand corporation (their advanced spacesuit prototype field test unit is not pressurized), let alone the Mars Society. Given this reality, we simply had to accept the limitations imposed by our finite means and use baggy tailoring to duplicate to a modest extent the movement restrictions imposed by puffed-up inflated spacesuit material.

Another important deviation from actual spacesuits was in the waste management system. Real spacesuits have pee tubes and diapers to deal with metabolic waste. For sanitary reasons, we decided to dispense with these systems, and installed a fly instead. It is breaking sim to use the fly, so it is almost never done, but it is good to have if you really need it. Women crewmembers carry a Lady Jane, a device developed for female athletes that allows members of the fairer sex to pee like a man.

There is no standard existing design today for a Mars spacesuit. The spacesuits used by NASA on the Space Shuttle today are designed for zero-gravity work in free space, and would not be functional on the surface of Mars. The Apollo spacesuits used during the 1969–72 period were lighter and more flexible, as they were built for planetary exploration. While the maintenance requirements of these suits would make them unsatisfactory for extended use on Mars, they are the closest examples we have to what a real Mars suit would need to be like. We therefore chose to model our suits loosely on the Apollo units.

One area where we chose to deviate from the Apollo suits, however, was in the helmet design. The Hamilton Sundstrand Mars concept suit we field-tested briefly on Devon Island in the summer of 2000 featured an improved visibility helmet for the user by providing him with an entire transparent hemispherical half-dome window. The advantages of this over a narrower transparent faceplate concept were so manifest that we decided to implement it in our suits as well.

The helmet of our simulator, like that of an actual spacesuit, does not turn with the user's face. Rather, it forms a bubble within which your head can move. Into this bubble we had to plumb air and communication lines. Air was supplied through two hoses by two computer fans located within a backpack unit that also carried the fan's rather heavy rechargeable batteries (sufficient for fifteen hours' use). The communication link was provided by an earpiece and microphone attached by wire to a walkie-talkie. The radio was positioned on the outside of the spacesuit so the user could adjust its channel and power levels. Both push-to-talk and voice-activated radio operation were possible.

The selection of a boot proved to be a major sticking point for a while. If all we had needed were a shoe garment that looked like a spacesuit boot, a Teva Moonboot would have filled the bill in superb fashion. Unfortunately, these items were found to have insufficient support to be used in field service on Devon Island. Hiking boots would have been quite serviceable, of course, but they did not look like spacesuit boots and made climbing much too easy on the user. In the end we found a satisfactory choice in the form of surplus U.S. Army cold-weather boots, the bulbous black rubber items sometimes known as Mickey Mouse boots for their resemblance to the footwear worn by the celebrated cartoon character. These turned out to be an excellent practical choice for Arctic service, as you could walk right through a freezing cold stream wearing them and not even feel it, and their imperviousness to touch and medium degree of clumsiness made them a good functional simulacrum for a spacesuit boot. When black gaiters were added, they had a bit of a spacesuit boot look about them as well. Finally, as an extra advantage they were cheap; we were able to pick up a whole load of them across the full range of sizes we might need from Front Range military surplus for $15 a pair.

To duplicate the Apollo look in a very Arctic-functional form, we chose to make the body of the suits out of white canvas. When a light spray of waterproofing was added, this stout material proved very functional under Arctic conditions. Cut to very large sizes to accommodate even the stoutest of crew members, these suits could be filled with sweaters or parkas to keep the user warm on a very cold day, while on hot days they had fair capacity to breathe. When subjected to field tests in Colorado, the suits were found to be functional at temperatures as low as 20°F and as high as 80°F.

The making and the sewing of the patterns for the suit bodies was led by Patty Piteau, an RMMS volunteer who is a seamstress and costumer by profession. There was a great deal of sewing to be done to create six working suits from scratch, and Patty couldn't do it all. So she recruited a sewing circle out of the bunch of engineers and scientists who constituted most of the RMMS, and taught them the craft. She taught them well; it's been two years since the suits were sent north, and despite some very rough service, they have held up like champs.

THE APRIL ADVANCE EXPEDITION

As the field season drew close in April 2001, we sent an advance team to the Arctic to check out the condition of the hab after the winter and to finish building out the interior. The team was led by Frank Schubert, and included Matt Smola, Matt's brother Len, a young Mars Society aerospace engineer named Greg Mungas, then working in my employ at Pioneer Astronautics, and Pascal. We chose April for this mission because the snow is still on the ground and you can land on the island in a Twin Otter, provided you replace the wheels on its landing gear with skis. This is not practical to do in May or most of June, as the snow is too thin then for the safe use of skis but still too omnipresent to allow wheeled landings until late June or early July.

So north they went, equipped with electrical wiring and plumbing materials, drywall, and conduit, to be joined in Resolute Bay by Inuk Ranger

Joe Amarualik. Loading their gear onto a couple of ski-equipped Twin Otters, they skied their planes in to encounter Devon's springtime weather of crystal blue skies, snow-covered landscape, and –40°F (which also equals –40°C) temperatures. They used skidoos to move their gear from the airstrip to Haynes Ridge, and warmed the well-insulated hab up nicely with the help of a portable propane heater. Frank, Matt, Len, Joe, and Greg spent a week working long hours fixing up the station's interior while Pascal, who is an avid photographer, walked around outside taking beautiful pictures of the snow. Pascal also put some labor in, which gave me reason to hope he might be moving toward embracing the team spirit that motivated other members of the project.

The final preparation for the summer 2001 simulation on Devon Island took place during early May, when twenty of the twenty-five selected crew members were brought to Colorado for training. The RMMS volunteers who had created the spacesuit simulators showed the crew members how to put on, take off, and use the somewhat complex suits, and also explained the reportage procedures they had worked out in the course of their mountain-excursion communication development missions. Then all the crew members were driven out to the mountains to conduct a twenty-four-hour mock mission. In the course of this activity, the crew members were divided into two oversized crews, one led by me and the other by Pascal, and each conducted two suited EVAs (extravehicular activities) to geologically interesting sites. Then, in the evening, each crew filed a full set of reports over the Internet with RMMS Mission Support, using RMMS communication protocols and with Earth-Mars time delays simulated in satisfactory fashion by having both Mission Support and the crews delay answering any received message by ten minutes. This practice field exercise was conducted in an atmosphere of good cheer, and overall was a complete success. By the time people left on Sunday afternoon, everyone was raring to go for the Arctic.

With the preliminary crew training done, the suits and other necessary simulation gear were no longer needed in Denver. As May drew to a close, the RMMS packed all their creations up in boxes and started the extended process of shipping them across the border to Edmonton and then from there via First Air to Resolute Bay.

Put-in on Devon Island was scheduled for June 25. This provided five days of preparations, during which Frank Schubert and some Inuit helpers would finish off the interior fitting out of the station. If all went according to plan, the first full-simulation Flashline Station crew would begin operations July 1.

THE MAKING OF THE MDRS

The spring of 2001 also saw another drama played out for our program, which was the creation of the Mars Desert Research Station.

We had found a good location in southern Utah for the MDRS, and in February 2001 an unusual opportunity arose to get it funded. Disney Epcot decided that it would be good idea to do an exhibition featuring the construction industry. A Tennessee-based exhibit impresario named Winston Huff, who ran a company called Science Interactive, convinced them that the most exciting way to do this would be to display construction's future—building cities in space. Huff then contacted Eligar Sadeh, a professor of space construction at the University of Colorado at Fort Collins, to get ideas on what should be displayed. Since we were the only ones who had ever built anything like a human extraterrestrial station, Sadeh put him in touch with us.

I proposed that instead of constructing the usual Disney theatrical mock-up, we should exhibit at Epcot the actual Mars Desert Research Station, ready for deployment to the field. This would have the advantage for Disney of making things more exciting for Epcot's guests, as they would be touring an actual ship of exploration. It would also have the more important advantage for us of proving the structure for our Mars Desert Research Station once the exhibit was over.

Disney liked the idea, but they weren't going to pay for it—nearly all the exhibits at Epcot are paid for by outside sponsors. Huff contacted several building trade unions, however, and intrigued them with the idea; an exhibit on space construction at Epcot would be exactly the sort of thing they needed to show young men considering a profession that the building trades were not vestiges of the past but the cutting edge of the future. Frank Schubert visited with the trades and convinced them that we had a team that could deliver the kind of exhibit they wanted within the time frame required by Disney. In the end, the United Association of Plumbers and Pipefitters (UA) and the International Sheetmetal Workers Union were sold; each would contribute $90,000 to sponsor the MDRS exhibit at Disney Epcot.

There was just one catch: Disney absolutely insisted the station be delivered to Epcot Center no later than May 31, as the exhibit was scheduled to open in mid-June, and it was already early March. We went to see John Kunz, but Infracomp could not guarantee delivery of a station on that schedule. So despite our appreciation for the quality of the product Infracomp had produced for Flashline Station, we were forced to look elsewhere for someone who could work faster for MDRS.

Searching fast, we found an alternative contractor in a Nevada-based firm called Built on Integrity (BOI), which was owned by Scott Fisher of the Fisher Space Pen family. BOI's technology was not as heavy-duty as Infracomp's. It basically consisted of a light steel frame that could be filled with polystyrene foam. The foam walls were then coated inside and out with an elastomeric resin, and decks were created by placing diamond-plate steel on the radial horizontal members of the frame. This system would not have been tough enough for use on Devon Island, but it would suffice for the milder conditions of the American desert. It was priced about the same as the Infracomp system had been and had the advantage of weighing less than half as much. This would make it much easier to assemble in the desert. Most important, however, BOI was willing to guarantee on-time delivery, and agreed to severe penalty clauses in our contract should they fail to do so.

On March 21, we signed the fabrication contract with BOI and gave them their first down payment to begin work. BOI immediately initiated procurement of materials and mobilized all their resources to meet the deadline.

Five days later, Disney canceled the exhibit. They explained me their reasons in a very polite phone call. Upper management had decided that space construction would be a more suitable theme for 2003. They were very sorry if they had inconvenienced us in any way.

Well, they had not only inconvenienced us, their action was threatening to bankrupt us. As soon as Disney canceled the exhibit, the building trades withdrew their monetary pledges. This was understandable and predictable; they had, after all committed their funds for an Epcot exhibit, not just a desert station. No show, no dough. Unfortunately, that left us on the hook with BOI for a very large sum of money that we no longer had.

I scrambled as fast as I could to find an alternative venue to recover the building trades support. I tried a number of prominent museums and exhibition centers, without success, then struck gold with a call to a friend of mine named Jim Ball. Jim was the former head of public outreach for Kennedy Space Center's Visitors Complex, the open-air rocket park display center affiliated with NASA Kennedy Space Center (KSC) at Cape Canaveral. He was also a founding member of the Mars Society. Jim said he would do what he could—and he certainly did. In a record-breaking eight days he ran the idea of displaying the MDRS at KSC right up the chain, obtaining the approval of the Public Relations Director and then the KSC Center Director himself.

Now the task remained to sell KSC to the building trades. Frank and I traveled to Washington, D.C., and met with their leadership. The union men were somewhat dubious. It seemed a poor substitute to them, as many more people visited Epcot than Cape Canaveral. I carefully explained the numbers to them. Yes, it was true that Epcot received 8 million visitors per year versus 2 million for KSC. Epcot, however, is so big that the average visitor only sees 20 percent of it. In comparison, the typical visitor to KSC Visitor Complex sees 70 percent of that facility. This nearly levels the playing field between the two venues, but KSC is better still, because its peak season is the summer, as compared with the school year for Epcot. Furthermore, something like half the guests at Epcot hail from foreign countries and are thus not prospective recruits for the U.S. building trades, while the large majority of KSC visitors are Americans. Thus, in point of fact, KSC was a better venue than Epcot.

George Bliss, the UA Vice President, who was attending the meeting, nodded. "We should do it. This will be our chance to show that you don't have to be a fighter pilot to be a space explorer." He asked if we could create an ancillary exhibit depicting cities being built on Mars—by union men. I said we could. He said the UA was in, and the Sheetmetal Worker representative quickly followed suit.

The return of union support saved our skins. It gave us the cash we needed to fabricate the MDRS and assemble it at KSC visitor's center. But we needed significantly more money if we were going to be able to staff it with Mars Society volunteer guides at KSC all summer, and even more important, deploy and operate it in the Utah desert the following winter. To help raise the necessary capital for this purpose, the Northern California Mars Society, led by NASA Ames human-factors scientist Bill Clancey and computer engineer Joel McKinnon, held a special fundraising dinner at Clancey's spacious Palo Alto home in early May. To help turn out the fading dot-com millionaires that still populated Silicon Valley, James Cameron volunteered to be the guest speaker of honor. The event was a success, and raised $45,000. Things turned out even better than that, because one of the people drawn to the event was a real heavy hitter, PayPal Internet bank founder Elon Musk. My wife Maggie and I arranged to meet Musk for breakfast the next day. That meeting went well, and he came out to visit with us in Colorado a week later, at which point he joined the Board of the Mars Society and donated a further $100,000. The full funding needed to build and operate the Mars Desert Research Station was now in place.

The deadline for delivery of the MDRS to KSC was the first week of June, offering only a few days' more latitude than had Disney's earlier deadline. BOI did manage to deliver the primary structure of the hab to KSC on time, except for one thing. A subcontractor was supposed to deliver direct to KSC a set of spacers that went between the wall panels, but failed to do so. I believe that BOI management was unaware of this problem. It now appears to me, however, that the man they had hired to oversee the job, and who was on site to assist Frank, Matt, and a group of Mars Society volunteers in erecting the station at KSC, knew all about it but chose to keep it secret.

KSC is not the Arctic. Construction equipment was available, and with the help of such aids, the lightweight MDRS walls were easily raised in a single day. But when Frank indicated it was now time to put the dome on top of the walls, the BOI man came up with all sorts of excuses for postponing work till the following day. Since it was already mid-afternoon, and the team had accomplished a lot by raising the walls so fast, Frank agreed.

The next morning, however, the BOI hireling still did not want to proceed. Before raising the dome, he said, they needed to get some left-handed widget-wacker from the hardware store. If they would just wait an hour, he would go get one. Frank did not understand why this particular tool was required, but if it were just a question of losing an hour to have the right tool on hand, he would do so. He told the BOI man to be back by 11 A.M.

Eleven o'clock rolled around and the BOI guy was nowhere to be found. Nor did he return by noon, or one. At two o'clock, Frank gave up waiting and decided to proceed with installing the dome without him. As soon as he did so, it became apparent why the BOI man had split. The dome panels were slightly wider at the base than the wall panels they were supposed to attach to. As they proceeded in installing one dome panel after another, the discrepancy accumulated. When the eleventh dome section was installed, the slice of the pie left open for the twelfth was a full foot narrower than it needed to be.

The cause of this misfit was the missing wall spacers, which would have added the necessary inch to the width of each wall panel to make it properly match its corresponding dome section. Frank didn't know this, and it wouldn't have done him any good if he did, since the spacers were never delivered anyway. So Frank being Frank, and Matt being Matt, they adopted the most obvious expedient and hacked a foot of width right off the final dome section to make it fit.

Now, in most respects doing construction at KSC is much easier than doing it in the high Arctic, but there is one additional complication: at KSC they have inspectors whose job it is to make sure that buildings are built in accord with certified plans. As luck would have it, one of these gentlemen happened to come along just as Matt and Frank were making their creative improvement in the dimensionality of the twelfth dome section. He ordered a stop-work until we could provide engineering calculations showing that the revised structure could withstand the hurricane wind loads it might be subjected to at KSC. This request would seem to be easy to answer, as the "revised" structure was not very different from the original, except that it turned out that BOI had not provided us with any certified calculations for that either.

Frank called me up at my office in Denver to give me the SOS: Find an engineer to do the structural calculations quick or we were sunk.

The request hit me at an inopportune time. Pioneer Astronautics was just wrapping up some key Phase 1 R&D projects for NASA, satisfactory performance on which it was necessary to secure the Phase 2 funding needed for the company to survive. The Mars Society had all sorts of packages en route to the Arctic, half of which were being held up in Canadian customs, or had been redirected from Edmonton to Cincinnati or Seattle, or anywhere but Resolute. *Newsweek* wanted to send its science editor Sharon Begley to Devon Island to cover our project, which would have given us great publicity, but Pascal was demanding the absurd fee of $8,000 to cover his expenses in hosting her and a photographer at the HMP camp for a week. (This proved irreparable. I forced Pascal to relent and offer them a more reasonable fee of $2,000—actual costs would have been about $1,000—but he told them it was a special discount. *Newsweek* refused the revised offer, saying their journalistic integrity would be compromised if they accepted such a subsidy!) I had a dozen crew members sending me urgent requests to arrange this or that, and my own affairs to set in order prior to going north as well. Still, an engineer had to be found.

I tried BOI first, as they claimed to have had a professor at the University of Nevada who had done the required calculations. But the material the professor sent me was inadequate, and would never have passed muster with the KSC inspectors. So I scrambled to track down some leads that Frank suggested, as they had done building-plan certification work for him in the past. One of these guys turned out to be available, so I rushed him the plans and then flew him down to KSC to inspect the structure and meet with the KSC building-code engineers. This did the trick, and the job was allowed to proceed.

The exhibit eventually proved to be a great success. Staffed by a group of Mars Society volunteers led by Pat and Tam Czarnik, it was used to explain Mars exploration to some 100,000 visitors over the ensuing summer.

Frank, however, had other responsibilities. Finishing the MDRS assembly project at Cape Canaveral on June 20, he flew back to Denver to spend a couple of days with his wife Olga, then headed north to hit Resolute Bay with our first wave on June 24.

The 2001 Arctic field season had begun.

8. LETTERS FROM MARS: DEVON ISLAND, 2001

Pascal, Frank Schubert, and the rest of the advance team reached Resolute Bay as planned on June 24, but a combination of bad weather and intermittent snow on the Devon Island airstrip kept them from proceeding farther for more than a week. On July 3, the weather finally broke and Pascal smartly seized the opportunity to rapidly shuttle eight Twin Otter flights into Devon loaded with people and supplies. On the fourth, however, the weather turned bad again, blocking the shipping of the remainder of our equipment until July 6. As a result, our first crew rotation, which was to have started at the end of June, only moved into the station July 7. On the same day, I reached Resolute.

Aziz met me at the airport and brought me to his hotel, where I encountered three members of my crew for the second rotation. These were Bill Clancey, the industrial psychologist from NASA Ames Research Center; Katy Quinn, an Australian-born geologist just finishing her Ph.D. at MIT; and Vladimir Pletser, a physicist and Belgian astronaut candidate working with the European Space Agency. I called a meeting, and explained to them that in view of the delay, the start of our rotation was being postponed from the morning of July 8 to the night of July 10. They were clearly disappointed by this turn of events. But everyone recognized that several members of the first crew, particularly Frank and Sam Burbank, had worked very hard over the past few days to get the station ready, and it would hardly be right for their shift to be stripped down to next to nothing.

Colleen Lenehan, a Canadian graduate student who was running logistics for the HMP in Resolute, sent us word that a Twin Otter would be available at 4 A.M. to take us all over to Devon. One of the crew demurred, saying he would prefer to go over on a later plane. In view of the uncertainty of the transport situation, I made it clear: when a plane is available, we would take it. Anticipating this early start, the crew turned in.

While the plane may have been scheduled for 4 A.M., it actually showed up at 12:45 A.M.. Colleen called me in the middle of the night to break the news. Get your team down to the airfield now, because the Otter is coming in and the pilot doesn't want to stick around. Everyone needed to be at the airfield by 1:20 A.M.

No one could find Vladimir Pletser, however. Colleen had phoned his room and knocked on his door, with no answer. Since there was nowhere else to look, I went down to join Colleen at his door and gave it a proper pounding. The door opened wide, and there stood a bleary-eyed Vladimir, dressed in the garments God gave him. Colleen, an Anglo-Canadian, appeared embarrassed, but the Belgian Vladimir was fazed not at all. I told him to throw on some clothes and hop in the van. Fortunately, Aziz had left the keys in the ignition. We barreled on down the dirt road and got to the field in time to meet the plane.

Twin Otters are small planes, and it did not seem possible that the little craft could contain our party, with its extensive baggage and scientific equipment, plus plenty of cargo that had been shipped up earlier by First Air to support the expedition. In addition to our own gear, we had eight boxes of spacesuit simulators from Denver, four crates of seismic sensing equipment sent by the French geophysical institute IPGP for Vladimir to use as part of the research program, and lots of other stuff. Miraculously, it all fit, and we took off at 2 A.M., into a brilliantly sunlit Arctic night. The sky was cloudless, and looking down we could survey the barren terrain of Resolute's Cornwallis Island, then the beautiful crazed patterns of the sea ice, and finally the primal Mars-like landscape of Devon Island itself. The plane began to descend, and then suddenly there was the Flashline Station, standing proudly on Haynes Ridge and flying the red, green, and blue Martian flag we had hoisted a year before. An instant later we were on the ground.

Landing at about 3:30 A.M., we were greeted by only two people, camp manager John Schutt and his deputy manager, Joe Amarualik. We unloaded the plane, pitched our tents, and sacked out.

My crew spent most of July 8 getting trained in ATV and shotgun use, and becoming somewhat familiar with the realities of life on Devon Island. On July 9, Bill, Katy, and Vladimir laid out the water pipeline from the Lowell Canal to the hab, while I went to the station with A. C. Schutt (John's wife) to clear away much of the construction debris from 2000 that still littered the area around the station.

When I got to the hab, I found Frank Schubert outside rigging up a spare 1.8-kilowatt generator. One of our two 5.5-kilowatt units had failed, leaving the station power strapped. The two largest power consumers in the hab are the incinerator toilet, followed by the water heating system. If the two are run together, the single 5.5-kilowatt generator can't handle it and throws its circuit breakers. I suggested installing a prominent switch and adopting the practice of turning off the water heater when the toilet is in use. Frank said he would do it. This decision was to put severe limits on hot-water use in Flashline for the season, but there was no choice. As a further backup, I spoke to John Schutt and set in motion the process of obtaining another 5.5-kilowatt unit before our last one failed.

Frank joined A.C. and me in moving away the debris. While working with Frank, I managed to get a more complete picture of the way the simulation as run by the first crew was going. Basically, it was a mixed picture.

Starting pains had made it necessary to break simulation constraints repeatedly. This was understandable, and in fact expected. The crew had been quite liberal in taking hot showers, however, which in my view was not realistic and was contributing to their power problem. Conserving water is essential to reduce the mass and cost of human Mars missions, so properly run sims needed be much more strict in this area. Instead of daily showers, future crews would be held to sponge baths every other day. The first crew also had its dinners cooked at base camp and sent over in plastic bags, purportedly under the belief that the meals of a Mars crew would be precooked and packaged on Earth, like MREs or TV dinners, to save crew time. I don't believe this. I think it would be very bad for morale for a crew of a 2.5-year Mars mission to eat MREs every day. The ritual of cooking real meals is an important part of human life. Accordingly, my teams would cook their own dinners.

On the other hand, Pascal was being quite strict about enforcing extended prebreathing as part of extravehicular activity (EVA). Under the assumption that the hab on Mars would be pressurized to 8 pounds per square inch (psi), he had specified a 30-minute prebreathing period prior to EVA in order to prevent decompression sickness when entering 4-psi spacesuits. Such a prebreathing period would indeed be necessary if a Mars habitat were pressurized at 8 psi, but it would impose a heavy impediment to the kind of frequent EVA an effective Mars mission would need. For this reason, I don't think a Mars mission should be conducted with a habitat pressure of 8 psi. Instead 5 psi (3 psi oxygen, 2 psi nitrogen) should be used, as it was on Skylab. If this is done, no prebreathing is necessary. Instead, the only delay during egress would be about a five-minute wait to allow for airlock pumpdown. In addition, the lower pressure habitat would require less structural mass than a high-pressure unit and would waste much less nitrogen through leakage or airlock cycling.

Around noon, Darlene Lim and Sam Burbank, two members of the first crew, conducted a pedestrian EVA along Haynes Ridge, to emplace some weather-logging instruments at its western edge. This was the second EVA the first crew undertook; the previous one had been undertaken by Frank and Sam the previous day in the immediate vicinity of the station to connect some sump disposal lines. On this occasion, Darlene and Sam took about an hour to walk a kilometer from one end of the Ridge to the other. They moved fairly slowly, looking down at the rocky ground all the time. Since Darlene is a paleontologist, I thought they might be fossil hunting, but it turned out they were just being careful. Nevertheless, it was the longest distance field trial of the suits on Devon Island so far, and its success relieved some fears that the awkward garments might not be safe to use under Arctic conditions.

After the EVA, Frank and I performed the ritual of replacing the worn and torn tricolor Martian flag that had flown over the station since the previous summer with a new one. We both climbed the antenna mast that runs up along the west side of the hab, and then as I managed the belaying lines, Frank climbed to the center of the dome and attached the flag to the pole.

Frank stood up in the center of the dome side by side with a mast on which a five-foot red, green, and blue flag of the future was now snapping proudly in a fine breeze, and took in the view. But he was the sight to see.

That evening the Rotation 1 crew conducted a motorized EVA, surveying some of Devon's canyon systems that closely mimic those found on Mars. Led by Pascal, and including Charles Cockell of the British Antarctic Survey, Rainer Effenhauser of NASA JSC, and Frank Schubert, the excursion really illustrated the unique capabilities for wide-ranging field exploration over unimproved terrain that human exploration teams will bring to Mars.

July 10 dawned sunny, although rather cold. As it was our last day operating out of the station, we decided to use the time to do a test deployment of Vladimir's seismic sensing array experiment. The device, known as a geophone flute, had been provided to Vladimir by the Institut de Physique du Globe de Paris. It consisted of an array of twenty-four sensors, which you need to stick into the ground in a line roughly 100 meters long. You then initiate a seismic signal, either with a sledgehammer, a gun, or an explosive. The sound goes into the ground and reflects off of various layers, and is then read by the sensor array. After sounding with the line in one direction, say north-south, you move the sensors 90 degrees to direct the line east-west. The combined soundings then produce a three-dimensional map of the subsurface. The purpose of such an array on Mars would be to search for underground water or ice.

Underground water is the most likely place to find actual extant life on Mars; reaching such water is thus the brass ring for Mars explorers. But even locating underground ice lenses in low-latitude regions of Mars would be of extraordinary value to a future Mars base. Mars is like the ocean; most of its secrets are to be found beneath its surface. For this reason, the ability of astronauts to deploy underground exploration equipment is key. We planned to attempt to demonstrate such techniques under simulation conditions during our upcoming crew rotation, but because of the complexity of the equipment, we thought it wise to attempt a test deployment out of sim first.

So Vladimir, Katy, and I loaded Vladimir's equipment into an ATV trailer and set off from base camp. We forded the Lowell Canal, which was still running high with ice-cold Arctic meltwater, and then scaled Haynes Ridge a few hundred yards west of the hab. It took about two hours to deploy the array, with some delay self-imposed as we made Vladimir do the wiring connections on the control box wearing thick ski gloves, since that is the way he would be impaired when we deployed the experiment under EVA simulation conditions. We strung out the array north-south, and then Katy used a sledgehammer to create a series of ten seismic signals. Amazingly, we got beautiful data from as deep as 300 meters on our very first try. Vladimir was delighted.

In addition to Vladimir, Bill, Katy, and me, the crew of Flashline Station Rotation 2 included British Antarctic Survey biologist Charles Cockell and Simon Frasier University electrical engineering Professor Steve Braham.

That night we moved into the station.

LETTERS FROM MARS

During the summer of 2001, I sent a series of daily dispatches describing the activities on the island. These appeared on MSNBC.com under the title "Letters from Mars." Selections from the dispatches tell the story of my two crew rotations.

JULY 11, 2001

Our crew moved into the Flashline Station shortly after 9 P.M. last night. I had hoped for a celebratory meeting of the two crews as they met during the exchange, but it didn't happen that way. Instead, the shortage of ATVs made it necessary for us to travel to the station two at a time, so that two of the first crew would return the vehicles for the next two of ours, and so forth. Vladimir and I got to the hab first. Sam Burbank explained to me the electrical wiring system of the hab, making it emphatically clear that the system was Frank Schubert's creation, not his own. I saw his point. The wiring system was clearly designed by someone with a unique sense of humor. Appropriate alterations are planned. We spent a bit of time stowing the large shipment of food we have brought with us, and then retired for the night.

In the morning I held a crew briefing laying out our goal of progressively increasing the fidelity of the simulation, and explaining how that will be achieved. Charles Cockell, our biologist, brought up the question of what we should do if an interesting scientific sample can be acquired by breaking simulation. I said we lose it. Our goal is to explore how much science we can accomplish under simulation constraints. We will break simulation for safety reasons, or to get a necessary piece of infrastructure, such as Steve Braham's advanced satellite communication dish system, into initial operation. Other than that, we accept the burdens of living and operating within simulation limits.

In a certain sense, the fact that everyone knows that in reality simulation constraints can be broken without risk puts an extra psychological burden on every member of the crew. A crew on Mars will know they need to act as if they are on Mars, or they will die. Here, we all know there would be no consequences if we go outside without spacesuits and do whatever we need to do the easy way. It is only the willpower of the crew that sustains the sim.

Following the briefing, we spent two hours cleaning up and straightening out the hab to put it in shipshape condition. Afterward, everyone was very glad we made the effort. Getting rid of the clutter practically doubled our working space and made the whole station much more pleasant.

The plan for the afternoon is a paleontology walking EVA on Haynes Ridge.

The crew selected for the EVA was Katy, Vladimir, and yours truly. Katy and I had worn the spacesuits before during training in Denver in May, but several changes had been made by the suit designers since, and it took about an hour to go through the checklist to get suited up. The suits are not real spacesuits but are quite elaborate nevertheless. Developed by a team from the Rocky Mountain Mars Society, led by Dewey Anderson, the suits look somewhat like Apollo gear and include an electric-powered air-ventilation system that is essential to the operator, voice-activated dual band UHF/VHF radios, and a water provision system with bite valves inside the helmets. The whole suit—including helmet, radio, backpack, fabric, water, batteries, air lines, boots, and so forth, weighs about 40 pounds. . . . When you wear one, you feel very much that you are moving around inside of a kind of bubble. You cannot touch or smell anything in the surrounding environment, and your hearing is greatly impaired. To talk to anyone more than 10 feet away you need to use the radios.

We are operating our sim under the assumption that the hab is pressurized at 5 psi (3 psi oxygen, 2 psi nitrogen, as on Skylab), which eliminates the need for prebreathing prior to EVA. So all we needed to do after the suits were donned and checked out was to spend 5 minutes in the airlock to allow for pumpdown before opening the door.

We proceeded west along the ridge from the hab. The ground consisted mostly of sharp rocks, and our vision was somewhat impaired by a slight drizzle, which wetted our helmets, so we had to walk carefully. After about 5 minutes I found something that looked like a fossil and showed it to Katy, who is a geologist. She confirmed my intuition. Then we all started to find additional examples of superior quality. The stuff was brain coral, probably dating from around 400 million years ago when Haynes Ridge was a reef basking in tropical seas.

A little while later Vladimir upturned a rock revealing an odd green discoloration on its underside. We suspected it might be a colony of extremophile cyanobacteria, and took a small sample.

As we proceeded farther west we crossed into what seemed to be a different geological unit of lighter-colored rocks that contained no visible fossils. So we doubled back and started exploring along the ridge to the east of the hab, where the rocks may be a bit older. We found more brain coral, then, eureka! Katy finds a stromatolite the size of a quarter. Then I found one the size of a baseball. Then Katy discerned a three-foot-square slab with half a dozen stromatolites embedded in it, a virtual mother lode.

This is terrific. Stromatolites are macroscopic fossils left by colonies of bacteria, exactly what we will be looking for on Mars. They are not as easy to find as conventional fossils of animals and plants, as they are much less distinctive. Yet we found them on Devon Island under suboptimal EVA conditions in less than two hours.

Could a robot landed on the ridge also have found the stromatolites? I believe the answer is no. Two hours is not a long time, but in that span the three of us wandered all over the ridge and our eyes processed the equivalent of millions of still images looking for very subtle clues. We could have wasted centuries trying to explore the ridge with robots and never made the discovery.

We returned to the station after two hours and gave our sample to Charles to analyze. Vladimir cooked us a great dinner of tuna fish, rice with tomato sauce, and vegetables. Eat your heart out, base camp.

Mission Support tells us that DHL says it has finally (after two weeks) delivered my water-testing gear to First Air for shipment north. We'll see. So far they've managed to ship it over 10,000 miles in various directions and circles around North America. This situation is beginning to remind me about the old song about the man who never returned because he couldn't get off the MTA. Call this one "The Gear That Never Arrived."

JULY 12, 2001

We held a morning meeting after breakfast and then set things in motion for a very involved three-person EVA.

The plan was to deploy Vladimir's geophone flute on Haynes Ridge under Mars mission EVA constraints and then fire it up to engage in subsurface seismic exploration. The team would be Vladimir, Katy, and I—the same group that had trained with the instrument while we were out of the simulation two days ago. The EVA would be quite complex, because the geophone flute consists of twenty-four sensors and a trigger placed into the ground at 4-meter intervals along a 100-meter line (it is this row of holes in a line that has given the system the "flute" name). All this would be wired together along with a computerized data-acquisition system that needs to be programmed in the field. Collectively the apparatus weighs over 300 pounds.

It took us about forty minutes to get suited up and out of the airlock. Then we loaded the equipment into an ATV trailer and trundled off across Haynes Ridge. The weather was cold, with a light drizzle that wetted the outside of our helmets and impaired visibility somewhat. Fortunately, we had taken care to soap the inside of our helmets, and so avoided fogging. Upon reaching the selected site, we first laid out a 100-meter tape measure in a roughly north-south direction. Then we scraped away rocks to push each of the twenty-four sensors about 6 inches into the ground. Following that, we unspooled heavy wire cables 50 meters in each direction from the center, after which we hooked up all the sensors using a series of connectors. The trigger was then emplaced, and all the wires rigged up to a battery and computer data system located near the sensor array midpoint. Then we turned the computer on. The power indicator glowed but no display showed.

This was a real problem. The geophone flute must be programmed with certain initial conditions each time it is set up, and it is also necessary for the computer operator to use data generated in real time to interact with the person causing the seismic signals if useful results are to be obtained. Neither of these essentials would be possible if the data display could not be read in the field. Peering at the display very closely I realized that the screen actually was turned on but was too faint to be read in outdoor light, especially given the added visual degradation caused by our wet helmets and the presence of rainwater on the computer screen. There was a brightness control in the computer, but it was digital, not analog. In order to be able to brighten the screen, you had to be able to read the screen. Catch 22.

I had some unprintable thoughts about engineers who design systems this way, and I believe my compatriots spent a moment in similar reflection. Then we set about finding a solution.

The answer we hit on was to pull the large foam panels out of the geophone's storage box and create a dark enclosure around Vladimir and the computer. Under these conditions he was able to make out the pale letters on the screen and increase the brightness and contrast to program the thing and proceed with the experiment.

Then we wired in the trigger phone, put a steel plate nearby, and stood back as Katy blasted it with the sledgehammer ten times. A beautiful pyramidal array of graphical data appeared on the screen, with sonic echoes from as deep as 550 meters!

We then moved the trigger and hammer plate 50 meters to the north end of the array, unspooled some wire to connect them to the data system, and repeated the experiment. Again we obtained good data. So we did it again at the south end of the string and were successful again. Then we pulled out all the sensors, disconnected everything, spooled in the wires and cables, stowed it all back in the boxes, loaded the trailer, and returned to Flashline Station. The total EVA time was approximately four hours.

Back in the hab, Vladimir analyzed the data. It showed no subsurface water or ice, which would be key targets for seismic searches on Mars. Instead, the seismic data supported the conjecture that the ridge is mostly simple dolomite rocks.

Three points, however, were forcefully driven home by the exercise. First, serious planetary research requires field operations that are far outside the existing or prospective range of capabilities of robotic rovers. No robot, or group of robots, could have done what we did today. Second, humans, even when impaired by spacesuit simulators, lack of tools compatible with spacesuits, bad weather, and a poorly engineered computer display system, can do whatever is required not merely to perform very complex operations but to actually make them succeed. Finally, something that was made clear to all the EVA's participants in a really sensuous way is that exploration is a very physical activity. The equipment we used was heavy, and it had to be lifted and carried significant distances over rough ground. At the end of four hours doing this kind of work in spacesuit simulators, all three of us were feeling pretty tuckered out.

This underlines an important point. The amount of exploration, both in quantity, quality, and variety, that a human crew will be able to accomplish on Mars will be directly proportional to their physical condition upon arrival.

Going to Mars in zero gravity is a bad idea.

JULY 13, 2001

It rained today.

Sometimes it was light rain, sometimes heavy. Sometimes it was just rain, at other times it was mixed with a bit of snow.

We hoped for a break in the rain so we could do a brief EVA to retrieve data stored in Bill Clancey's Campbell Scientific weather station, which is positioned near the airstrip across the valley. We actually got a brief period when the rain was light enough to try it, but at that time the Lowell Canal, which cuts through the valley, was running so high that a person crossing it on an ATV would get soaked up to his hips.

So it was an indoor day. Charles Cockell made good use of the time to analyze biological samples gathered by our Haynes Ridge EVA team July 11. It turns out that the green discoloration Katy spotted on the underside of some rocks was indeed cyanobacteria of a type typically found in extreme polar deserts. It grows on the edges of the rocks, just underneath the surface, where it is protected from UV radiation and can live in its own microenvironment. Enough light penetrates around the edges of the rock for these hardy microbes to photosynthesize. Examining the material with our epifluorescence microscope, Charles was able to identify filaments and spherical microorganisms and photograph them under near-UV light for further study.

The rest of us used the in-station day to catch up on our writing.

Our power problem persists. The station was designed to operate on 10 kilowatts. Since the failure of one of our two main generators, we actually have 7. So if someone turns on the lights downstairs when the water heater is running, it throws the circuit breaker. It is equally unwise to run the microwave and the hot plate at the same time. We don't dare to turn on the space heaters, so the station internal temperature varies between 50 and 60°F. Charles and Steve, who have both been here since the start of the first rotation, really needed a Navy shower, and wanted to do it with warm water. We had to delay lunch so as not to use any cooking power while the water heater was running. We get by watching every watt.

Being power-poor is a drag. A human Mars mission better bring a nuke.

JULY 14, 2001

Last night we finally were able to get everyone to stop working at the same time and have some group R&R watching a DVD projector image of the movie *The Vertical Limit*. If you've seen this movie, you know that the scenery and cinematography are incredible, but the action is ridiculously contrived, with the climbers making one absurd mistake after another in order to maintain the momentum of the film as an avalanche of literal cliffhangers. It had us all in stitches.

The fun was interrupted, however, by news of a serious medical problem at camp, which we picked up over our UHF link. One of the visiting journalists started suffering from dizziness and shortness of breath. Rainer Effenhauser, who was a member of the first crew, is a NASA flight surgeon, but he is back at Johnson Space Center now. We could hear over the radio as he directed Pascal by satellite phone to take the man's blood pressure and conduct other parts of a medical examination. This was telemedicine, done not in simulation but for real; it may also be the way medicine will need to be conducted on a mission to Mars. The decision reached was to pull the man out. He was evacuated by Twin Otter to Resolute this afternoon.

The plan for today was a four-person motorized EVA to Trinity Lake and Breccia Hill. Trinity is located in the crater, and so we had serious concerns over whether the ground was dry enough to get there by ATV. The mission was led by Charles, and included Katy, Vladimir, and Bill Clancey. Its purpose was to place cosmic ray dosimeters in both of those locations and to collect samples of rock-dwelling microbes for Charles's analysis back in the lab. Placement of such dosimeters will be a task necessary on Mars, and is of significant interest here, since at only 140 miles from the magnetic pole, Devon is a locus where charged particle radiation from space is somewhat focused. The EVA lasted three hours and was entirely successful. The new high-gain backpack antennas worked well, allowing the EVA team to communicate back to the hab from 2 kilometers away with their dual-band transceivers on low power. This was quite fortunate, because for some reason, when put on high power, the things run down their batteries here in a time much shorter than they do back in the lower forty-eight. I've noticed the same behavior from my digital camera. In thirty shots, the batteries are gone. It must be the constant cold.

Anyway, as things turned out, the crew was able to make it to Trinity without having to dismount from their ATVs and hike. This allowed the EVA to accomplish its primary objectives ahead of schedule, so we gave them a go-ahead to conduct some auxiliary reconnaissance to start searching for a scientifically interesting site to deploy Vladimir's geophone flute for another subsurface sounding experiment. They found a place within the crater near two converging streams that seems to be of some interest, but we are hoping for better.

I received a message through Mission Support from my wife Maggie telling me about happenings back home. Our older daughter Sarah, eighteen, is going down to KSC to help exhibit the soon to be deployed Mars Desert Research Station. Our younger daughter, Rachel, age nine and away at summer camp in Colorado, has just climbed a 14,000-foot mountain. They both make me very proud.

Messages from home will play an important role in any Mars mission. If the news is good, it can be an important boost to morale. But if the news is bad, the impact could be devastating. Figuring out how to handle such messages could turn out to be one of Mission Control's most important tasks.

JULY 15, 2001

The weather today was fair, with the prospect for better tomorrow and Tuesday. So we decided to seize the time to perform a reconnaissance EVA to try to find the best site for Vladimir's geophone flute seismic sounding experiment. The crew was Vladimir, Katy, and I, and our target area to survey was the Von Braun Planitia, a large, flat expanse stretching north from the hills overlooking the Lowell Canal.

Our method of transport was ATVs. We saddled up, and set course for Von Braun. To get there, we had to descend from Haynes Ridge and cross the Lowell Canal, which was running fast this morning with crystal clear ice-cold water about 18 inches deep. Then we forded more streams and climbed the slopes to reach the plateau. We then drove several miles across the plain, looking for a geophone site. Some of it was quite muddy from all the rain and snowmelt that had been inundating the area until recently, and we had to keep the vehicles moving to prevent sinking.

We made rapid progress traversing to the edge of the planitia, after which we crossed a stream and scaled a rocky ridge. Here the terrain changed into a small canyon system. The area impressed me as being interesting to explore, but not useful for geophone purposes. As we had already been out for over two hours, we decided to head back.

Polar bears have been spotted on Von Braun, and spacesuited crew members on EVA lack the situational awareness required to be safe from them. So on this EVA, as on all others involving travel significant distances from the station, the crew was followed by an armed Inuk from camp. In our case, it was Joe Amarualik. Joe, who also serves as Deputy Camp Manager, is an experienced ranger with many bear kills to his credit. He carries a Lee-Enfield .303.

Upon returning to the hab, we held a general meeting of the crew to discuss the comparative merits of the various sites visited. Our conclusion was to try the site in the crater where two streams meet, about halfway between the station and Trinity Lake. There may not be ground ice there, but at least we will be able to make a useful measurement, sounding down through the breccia left by the meteor impact to determine the depth of the crater floor beneath it.

There is a shower in the hab, but to save water we don't all use it every day. Instead, two people take either water-conserving Navy showers (in which you turn on the water just long enough to get wet, then soap up, then rinse off) or sponge baths every day, which means that each crew member gets to wash every third day. Today was the turn for Bill Clancey and me. We both elected sponge baths, and averaged about two gallons each. This is much better than the Navy showers, which according to our records average about 6 gallons. We had the water heater on for the occasion, but as I went first, the hot water did not make it down the line in time for me to use. Cold sponge baths are very refreshing. Try one.

JULY 16, 2001

We took some time off last night to watch another movie. This time the selection was the alien invasion spoof *Mars Attacks.* The crew got a real kick out of it. In fact, they got so high on the film that this morning, when the EVA team did their radio voice checks, everyone substituted Martian talk ("Ack, Ack Ack, Ack Ack Ack") for the usual integer count to ten.

What followed, I confess, was less than fun. The mission was to transport Vladimir's geophone flute into Haughton Crater, to the site we call Two-Streams, and set it up twice to get a complete three-dimensional seismic sounding. The EVA team was made up of Charles, Vladimir, Bill, and me, with Katy running the capcom station communication link supporting us from the hab. The Two-Streams site is only a couple of kilometers from the station, but to get there we had to descend from rocky Haynes Ridge into the crater, whose surface is covered with a combination of condensed rock vapor powder, which rained down after the impact, and eroded silt transported into the crater bowl by water action since. This fine material had been turned into thick, viscous mud by the recent snowmelt and rain, but during our sortie to Trinity Lake two days ago, we had been able to cross it by ATV without too much difficulty.

On this deployment, however, we were dragging the geophone — 300 pounds of equipment hauled by ATV trailer. That changed things considerably. Two-thirds of the way to the deployment site, the trailer and Vladimir's ATV hauling it got hopelessly stuck in the mud.

There was a bar of firmer ground about 20 meters beyond the mud trap, and we parked the other ATVs there. We got a rope and tied it to the trailer hitch on one of the free ATVs, then walked back to attach it to Vladimir's to give him a tow. As we approached the trapped vehicle, the quality of the ground turned to mush and we began to sink a foot or more into the mud with each step. The stuff would have been a deadly kind of quickmud, except that about 18 inches down was a floor of permafrost that limited the depth to which we could sink. But each step became a labor. Sometimes you had to dig the mud away from around your boot just to be able to lift it for another step. Our spacesuit boots are actually rubberized U.S. Army cold-weather boots, and they lace on pretty tight. Nevertheless, the mud gripped them so hard that several times I had to stop lifting and put my foot back down and dig around it in order to stop the mud from pulling a boot right off my foot. When you pushed your foot down all the way, you could feel the cold permafrost, and I had eerie thoughts about what might happen if the warmth of my boot allowed it to melt its way into the permafrost a bit, only to be frozen in afterward. So I did my best to keep moving.

We reached Vladimir's vehicle with the rope, but it was hopeless to try to pull his ATV and the trailer out at the same time. So we tied the rope to his front axle and detached the trailer from the hitch. Then Charles went back to drive the tow ATV, and with Vladimir, Bill, and me pushing, and running the stuck ATV to help out, we pulled the trapped vehicle out of the muck.

Now the task was to get the trailer out. This proved harder, as with only two wheels to support it, the trailer had sunk pretty deep. Tying the towrope to it, we tried to pull it out with an ATV, but it wouldn't budge. Charles, Bill, and I wanted to unload the trailer, drag it out, and then try to slide the three heavy-duty plastic crates containing the geophone out over the mud with an ATV pull. Vladimir, who was responsible for the gear, wouldn't hear of it. So Charles appealed for advice from John Schutt, the camp manager and experienced Arctic scout, who was riding shotgun for our EVA. This was a violation of simulation, but the situation was serious enough to merit such a move.

Schutt recommended we rig up a pair of ropes and try to pull the loaded trailer out with two ATVs. This we did, and with Vladimir, Bill, and me pushing from behind, and John and Charles gunning the two vehicles, we managed to haul the trailer free. The job had taken an hour and a half of very hard work, with all of us except Schutt doing it in complete EVA gear.

By this time, we had all agreed that the mission into the crater had to be aborted, so we started to head back. But we had not gotten 200 yards when the trailer and its tow ATV got stuck again. My heart sank. We now had to do the whole procedure all over. Fortunately, this time the ground was a bit firmer and we all knew what to do, so it only took about forty minutes to get them free. We then continued back, and caked with mud, we made it to the Flashline Station without further incident.

There is no mud on Mars, but there could be vehicle sand traps. So something like what we faced today on Devon could happen there. If it had, and the Mars expedition performed no better and no worse than we did today, they would not have lost crew (since we were able to get the ATVs out of the muck ourselves), but might have had to abandon equipment (since we had to break simulation to recover the trailer).

Vladimir is very disappointed with the day, since no geophone data was acquired, but I am not. We are here to learn lessons, and we received plenty of those. The two most important center on the basic issues: How did we get into trouble and how did we get out of it?

We got into trouble because our preliminary reconnaissance of the questionable crater terrain was too superficial. Yes, we drove over it in advance, but only in ATVs. We should have done more, perhaps taken a trailer down there, loaded with rocks to match the geophone weight. If that had gotten stuck, we could have just tossed out the rocks and returned. On the other hand, making it to the site with the rock-loaded trailer would be proof positive that the mission was doable, and commitment of the geophone would have been justified.

So much for how we got into trouble. Now, how did we get out? We got out with the help of John Schutt, but not because of any extra muscle or equipment that he added, or because he could operate outside of simulation burdens. Indeed, even after Schutt came to our aid, nearly all the labor was still performed by us, and those things that he did do, any of us could have done. The key thing that Schutt brought along that allowed us to save the equipment was his intimate knowledge of using ATVs under difficult Arctic conditions.

Knowledge is power. Mars explorers will need to train with their rovers until they know their vehicles' capabilities as well as John Schutt knows his.

JULY 17, 2001

After yesterday's brutal EVA, I decided to set an easier schedule for today. We spent the morning filling out psychological questionnaires and rejuvenating equipment impacted by our recent battle with the mud. Only after lunch would we attempt an EVA, whose mission would be to complete Vladimir's geophone survey of Haynes Ridge.

The questionnaires were sent to us by human-factors scientists from the University of Quebec at Hull (UQAH) and NASA Johnson Space Center. Some of the crew members object to the forms, because many of the questions appear trite. It is obvious that to develop a real understanding of the expedition, the human-factors types need to do more than send questionnaires; they need to go into the field themselves. To be fair, the leader of one of the two groups, Judith Lapierre of the UQAH, proposed to do exactly that. She was among the four hundred people who applied to be members of the Flashline Crew when our call for volunteers went out last November, and in fact was one of twenty-five finalists before the final downselect forced us to choose just twelve from the volunteer group. Undeterred, Judith signed up to work full-time for no pay in Mission Support in Denver for the summer—at least she could observe us from that position. (She could also observe close-up the folks at Mission Support, who are in a very real sense vital members of the crew too, and thus equally worthy of study.) People as dedicated as Judith deserve to get their data.

The mud that encrusted all of our gear had dried by mid-morning, and we take stiff brushes to everything that can be brushed, and wash everything that can be washed. By 1 P.M., most of our EVA equipment is functional again. The suits are no longer a nice clean white, but they won't stay that way on Mars for long either. Around 2 P.M. we go EVA.

The afternoon weather is splendid: 50°F, sunny, and no winds; we dress lightly under our spacesuits. Even so, the helmets act like greenhouses, and it is not long before Katy, Vladimir, and I are sweating profusely. The geophone gear is not as well ordered as it was when it was first delivered from France, and we are faced with various problems, such as wires coming loose off of spools. But it is a job we are now well trained for and we manage to deploy the geophone in a line perpendicular to the position we sounded from several days ago. This gives Vladimir enough information to create a complete three-dimensional picture of the subsurface of Haynes Ridge down to 1600 feet, with all data having been acquired under simulation conditions. He is very happy.

The rotation between the second and third crews was set to occur at 9 P.M. tonight. I will stay on as commander, and Steve as engineer, but Katy, Vladimir, Charles, and Bill will be replaced by a new crew consisting of the French/American geologist and author Charles Frankel; Cathrine Frandsen, a physicist and planetary scientist from the Niels Bohr Institute in Denmark; Christine Jayarajah, a chemist from the University of Toronto; and Brent Bos, a planetary science graduate student from the University of Arizona. The new crew members are already at camp and I communicate my wish list to them. The food larder in the hab has been converging toward Slim Jims, dried fruit, and Tang. I suggest the incoming crew members bring some better stuff with them from camp. They manage to fill a cooler with canned fish, bread, pasta and pasta sauce, and other essentials. Well done.

The crew change takes place at nine as planned. Some of the new arrivals know the departing members from crew training in Denver, and many warm greetings are exchanged. The two crews work together to rapidly move the departing crew's gear out of the station and move the arrivals' material in. I shake hands with all the departing men and give Katy a hug.

Finally, the second crew leaves, and we, the members of the third crew, are alone in the station. I give the new team a preliminary briefing and ask them about the instruments they are bringing. Cathrine has a magnetic dust collector, Brent a sophisticated camera, and Christine an analytical chemistry set.

I tell Christine that I have an "instrument" for her. She does not understand. So I go up to the storage loft above the bunks and come down with an electrical keyboard. Her eyes light up. You see, I had information from other sources that Christine is a first-class pianist. She says she is out of practice and this and that, but we all urge her to give it a try. So she steps up to the keyboard, and fills the air with a Mozart Sonata in D Major.

Mozart on Mars!

JULY 18, 2001

The new crew awoke bright and early, and after breakfast and a short briefing, began to suit up for their first EVA. Because it would be their first sortie, I kept the plan conservative. Three crew members would do a two-hour pedestrian excursion on Hayes Ridge, to continue the general geological and paleontological survey of the area begun by Vladimir, Katy, and me in our first EVA a week ago. The EVA crew, consisting of Charles Frankel, Cathrine Frandsen, and Christine Jayarajah, would be led by Charles, with Brent Bos serving as their capcom in the station.

It was good we started early, because the EVA preparation took three hours, due to a combination of the new crew's inexperience and various technical problems, mostly radio-related. So it was 1:30 P.M. before they were out the lock. But if the preparation phase was a comedy of awkwardness mirroring the experience of the first efforts of prior crews, the actual EVA was superb. The weather was sunny, and the crew took full advantage of it to not only survey the ridge, but to clamber all over the boulder-covered steep slopes leading down into the crater. In doing so, they demonstrated the ability of human explorers to cope with a type of terrain that would be impassable for any wheeled vehicles or other robots.

Their science return was high as well. In addition to the paleontology, the crew also returned mineralogical information, and took rock, soil, and microbial samples for analysis back in our lab. They also deployed Cathrine's experiment, a Niels Bohr Institute dust magnetic properties instrument similar to that used on the Mars Pathfinder mission.

Dinnertime. At Flashline Station, dinner is a major group event, with attendance mandatory. Everyone takes a turn cooking. Going into the field season, I had imagined that cooking for the group would be a chore everyone would try to avoid until the fatal day arrived when he or she drew the short straw. But that is not how it has turned out. Instead, during the second rotation, a kind of friendly competition developed, with each cook attempting to outdo the others by creating an interesting dinner out of our stockpile of canned and dried goods, or at least trying to avoid public disgrace.

Working with our supply of room-temperature preservables, Vladimir rendered a meal of tuna fish, rice, and beans; Steve Braham cooked Chicken Alfredo (using canned chicken); Katy made beef stew and potatoes (using canned beef and powdered potatoes); I used the leftover beef from Katy's stew to serve spaghetti with meat sauce; while Charles Cockell prepared chili and canned vegetables (he's British).

The first member of the new crew to volunteer to cook was Charles Frankel, a dual-citizen Franco-American who has lived most of his life in France. I am a man of simple tastes, one who would happily cross the street to avoid a French restaurant. I therefore awaited his production with some trepidation. On the menu, however, was a hearty dish called Chicken Provence, made with canned chicken, canned tomatoes, rice, olive oil, and spices. It turns out that those absurdly precious meals consisting of a quarter-ounce of lamb topped by a twist of broccoli with some weird sauce on the side, etc., served by irritable waiters at the establishments calling themselves French restaurants are a kind of practical joke the French like to play on Americans. No one in France eats that stuff. They eat real food, just like everyone else.

This means that a joint U.S.-French mission to Mars is potentially possible.

JULY 19, 2001

This morning I outlined the campaign for the third crew. Unlike all the other rotations, this crew has no one in it with extensive Devon Island field exploration experience. For us, as for a crew newly landed on Mars, the surrounding territory is terra incognita. We have aerial photographs and a topographic map, but they do not tell us what the land surrounding us is really like, or what we will find there. So we must create a map with content. We must explore.

For the next several days, we will conduct a systematic survey over increasing distances in all directions around the hab. Our survey began yesterday, with the pedestrian reconnaissance of our landing site on Haynes Ridge. Today we would go farther, using ATVs to explore the hills and valleys to the east of the station. The EVA crew would be made up of Charles, Brent, and me, with Cathrine serving as capcom. Simultaneously with our excursion, Christine would begin her chemical analysis in the lab.

Our suit-up period was still slow, but at two hours compared to yesterday's three, it was a distinct improvement, and I anticipate better tomorrow.

Ours was a joyous excursion. The day was brilliant, with temperatures about 50°F and hardly a cloud in the sky. The hills to the east of the hab overlook the crater, and the views were spectacular. Charles collected rock, soil, and water samples, and Brent noted waypoint positions for sample collection in his GPS, communicating this data to Cathrine back at the hab. (Because it is very difficult to write while wearing a spacesuit, it is the station capcom who takes notes for the EVA team.)

My responsibilities were the lightest. All I had to do was indicate the direction of travel, take photographs, and tell everyone when it was time to move on.

Some of the valleys we passed through were still muddy and quite treacherous; as a result of the dry sunny weather we have had for the past two days, the mud is now covered by a thin layer of bone-dry dirt.

It is now stealth mud: you can't see it until you are in it. We had a couple of close calls, but we were traveling light and never got stuck more than momentarily.

We came to a hill overlooking a stream, and decided we would go down to it to get a water sample for Christine. Having by this time discovered the deceitful nature of the low ground, we decided to leave the ATVs on the crest and travel down on foot. The distance to be walked was 200 yards, and easy enough going down. The hike up in spacesuit simulators was a bit of a workout. But while it turned out that the ground we covered probably was strong enough for ATVs, reaching the idyllic locale at the bottom of the slope on foot made the experience somehow much more special and well worth the effort.

Then, sadly, it was time to go back. We had to return to the hab by 4:30 so we could interact with the public at the Mars Society exhibit at Kennedy Space Center, Florida. It turned out that our haste was for naught, since the KSC exhibit center was evacuated this afternoon because of a hurricane. Floridians should consider relocating to Devon Island. The weather here is fine.

Even though we were rushed, I saw something on my way back that made me slam on the brakes of my ATV. It was a flower.

It was like suddenly meeting an old friend in a strange place.

Then I saw a few more flowers, typically separated from each other by 100 meters or so.

Now this is a very interesting thing. Given that these flowers can survive up here, the question is: how come they are so few and far between? They are clearly too far apart to be competing with each other, and there are no insects or herbivores to prey upon them. So why is one here, and another way back there, and nothing betwixt?

The answer, I believe, has to do with the lack of animals. The meteor impact flashed to vapor all the nitrates in the soil in and around the crater, leaving the ground infertile. So the only places plants can grow are places where the few animals present have enriched. There is a flower at this particular spot, because thirty years ago a gull took a dump here.

A lot of people view the Earth's ecology as being built upon the productivity of plants, that are consumed by herbivores, and the herbivores by carnivores, etc., in a kind of pyramid, with the plants at the bottom supporting the whole nonproductive lot above. But this is a mistake. You can see it on Devon Island. Animals are needed to open new lands for plants.

Chipmunks transport pine cones up mountainsides, spreading forests high up the slopes. Birds drop fertilizer and seeds on newly formed sterile volcanic islands, transforming them into lush tropical oases. Animals create the habitats in which they and their posterity can live.

Mars is a currently barren island across the ocean of space. We are the birds that can bring the seeds of life to it.

Come, let us fly.

JULY 20, 2001

Today was the twenty-fifth anniversary of humanity's first successful attempt to land a robotic probe on Mars. On this day in 1976, the Viking 1 spacecraft set down on Chryse Planitia. It was also the anniversary of another occasion. On this day in 1969, Neil Armstrong and Buzz Aldrin landed their Apollo 11 spacecraft on the Moon.

It would be a disservice to human memory to allow an occasion of this magnitude to pass unnoticed and uncelebrated. What to do? We wrestled with ideas. Finally we hit on a good one. We sent our EVA team roaring through the HMP base camp on ATVs, like a bunch of Hell's Angels, flagging signs reading, in French and English, "Happy Anniversary, Viking," and "Viking, We Are Here!" It was a lot of fun. We took the whole camp by surprise, and CNN and Discovery Channel turned out to film the ruckus. Then the team, consisting of Charles, Brent, and me, sped off toward the Von Braun Planitia to do some science.

On our way north across the barren plain, we were greeted by a strange sight—the Hyperion robot. The Carnegie Mellon University machine had passed its initial tests with flying colors but was dormant now, just as Viking 1 is. We approached it gingerly. As we did, the thought struck me: someday a team of astronauts traveling across Mars will also encounter a robot standing motionless in the middle of nowhere. Perhaps it will be Viking 1.

I used to work for the Martin Marietta Astronautics company, which built the Viking spacecraft and helped fly the mission. I was not there when the mission was flown, but I was a member of the unit that had done the job, and many of my coworkers had worked on it. For most of them, Viking was the proudest and most profound experience of their professional lives. I have heard many Viking stories and feel I kind of know about it in the same sort of way I know about my uncle who died fighting in France eight years before I was born. Hence, encountering Hyperion frozen in place out on the planitia was more than a little eerie. I felt like I was meeting a ghost.

We traveled across the plain to a large rock formation, which towered perhaps 100 feet above the surrounding level plain. It contained exposed bedrock, which is what Charles was looking for, so we stopped to explore for a while. Poking among the stones, I came across a rock with an oddly curved back and wavy striations along its edges. I pulled it out and called out to Charles for his opinion. No doubt about it; it was a stromatolite, and a big one too. Stromatolites are the most likely form of fossilized evidence of past life that we can hope to find on the surface of Mars. Once again, one of our EVA sorties had managed to find them on Devon.

Then, while Charles collected rock, soil, and water samples for Christine to analyze back at the hab, I climbed the rock formation. It was only a 100-foot gain, but the way was steep and covered with large sharp rocks. No robot could have climbed it, but I managed well enough in a spacesuit simulator. The climb was worth the effort. Von Braun is flat enough that the elevation gave me a magnificent view, extending at least 10 miles in almost every direction. I called to Brent and Charles to join me, and we spent some time exploring the top of our little peak.

It should be noted that the ability of a human explorer to climb boulder-covered slopes to attain an elevated position has more than aesthetic value; it offers real scientific payoff. A camera that can be positioned on top of even a 100-foot peak by human explorers will return far more information than a robotic eye moving along slowly a foot or two above the ground.

Later, returning to the hab, we stopped to monitor the status of Cathrine's magnetic dust properties experiment. Then we entered, debriefed, and composed, taped, and transmitted a report to Mission Support before settling down to dinner. Tonight it was Christine's turn to cook. The menu was mixed (canned) salmon with (canned) spinach and rice. I don't know what to call the stuff, but it was pretty good.

I have been thinking all evening about this Viking and Apollo anniversary business. It is true that we must celebrate the great deeds of the past, but this celebration has a bitter taste. After all, it is tragic that a generation after these missions, no human being has ever flown farther than Apollo and no more capable spacecraft than Viking has ever landed on Mars.

Memorial Day was originally created to honor the Union victory in the Civil War, and Memorial Day 1965 was its hundredth anniversary. On that occasion, as a lad of thirteen, I witnessed a veteran's parade honoring the event. Of course the vets I saw marching were not Civil War veterans. By and large they were World War II veterans, and while honoring the hundredth anniversary of the victory of Union arms, they were also celebrating the twentieth anniversary of their own triumph.

It would have been a very sad thing if, one hundred years later, the most recent victory the U.S. Army could claim was Appomattox.

July 20, 2026 will be the fiftieth anniversary of Viking 1. I'll be seventy-four years old then, so there's a good chance that I will still be alive on that day. But I don't want to spend it honoring the fiftieth anniversary of Viking. I want to spend it celebrating the fifteenth anniversary of the first human footsteps on the Red Planet.

JULY 21, 2001

Explorers need maps. We have some good aerial photographs of the region around Flashline Station, but an aerial photograph is not a map. A map requires mathematical organization, which is imparted to it by a system of coordinates.

In most countries around the world, people use latitude and longitude to grid their maps. But in the Arctic, the distances represented by degrees of latitude and longitude differ immensely, and longitude degrees on the same chart can differ significantly from one another, making this system inconvenient. Instead, many people have come to use a system of grids known as Universal Transverse Mercator, or UTM. In the UTM system, locations are given in kilometer distances north of the equator and east or west of various meridians. In the UTM system, the location of Flashline Station is around 8379.7 kilometers N, 420.5 kilometers E.

We have a topographic contour chart with UTM coordinates in the station, but aerial photos give a much better sense of the nature of the ground. So in order to turn our aerial images into maps, this morning I used the data in the contour plots to lay down UTM grids on the photos, marking them off with 2-kilometer separations between lines. Our Global Positioning System (GPS) receivers have an option to display in UTM, and using it our EVA teams can report their UTM coordinates via radio. This in turn makes it easy for the capcom at the station both to follow the team's progress and to mark off their ground-truth characterizations of various features imaged on the photos. This latter function—ground truth—is very important. You can make out certain patterns on aerial photographs, but until you actually go in person, you don't know what they represent. After visiting a site, however, you have a pretty good basis for guessing what similar features in the image might be.

It is unlikely that an early human Mars expedition could have navigational assistance from a GPS-like space-based navigation system, as placing such an elaborate satellite constellation into orbit around Mars would cost much more than it would be worth. A cheaper solution would be to simply place radio beacons in several locations in the region surrounding the base. This would create a local navigation system, whose output could be read and analyzed to produce the equivalent of UTM GPS.

Our EVA team today was comprised of Charles, Brent, and Cathrine; their mission to explore the ridges and canyons to the north of the Von Braun Planitia. I served as capcom, while Christine continued her analysis work in the lab. Every fifteen minutes or so the team would call in with the UTM coordinates and site descriptions, and I would chart their progress on the map. They departed the station at 12:21 P.M. Slightly more than an hour later, they were almost 7 kilometers from the Flashline Station. Nevertheless, I could receive their handheld radios loud and clear.

I received several more reports from the team, but after 2:15 their transmissions ended. I tried to raise them every half hour, but there was no response. This did not worry me excessively; the team was headed for an area of ridges and canyons, and terrain could easily cut off their VHF radios. Moreover, the weather was fine and there were three of them to assist one another, plus Joe Amarualik riding shotgun.

By late afternoon I began to get a bit concerned about the EVA crew, but at 5:10 Cathrine finally reported in. The problem had been a terrain cutoff. All was well, but the team had had its share of adventure.

I got the full story when they reached the hab about 6 P.M. The crew had indeed descended the far side of a ridge to try to make it to a canyon but had repeatedly been blocked in their forward progress by various combinations of boulder fields, rocky moguls, and snow. Rather than give up, they had doggedly doubled back again and again to try alternative forward routes. They got stuck many times, but managed to help one another get free. Then Cathrine was thrown from her ATV, after which the vehicle rolled over her leg. This caused great concern among members of the EVA team, but intrepid Cathrine just kneaded her leg for several minutes and then insisted they proceed with the mission. [Note by RZ, December 2002: This latter detail was excised from the original dispatch in order to avoid a freak-out by the HMP.] So on they went, finally making it to a site where the canyon could be imaged. They took some rock, soil, and water samples, photographed everything in sight, and headed back, only to discover that the difficult terrain they had managed to traverse going downhill was much worse going up. Nevertheless, with suit batteries running low, they made it out and back to the hab without breaking sim. At close to six hours in duration, it was our longest EVA yet.

The third crew has the stuff.

JULY 22, 2001

After yesterday's rough EVA I thought it best to take it easy today. In any case, it was Sunday, and so we rested. Instead of going EVA, we spent the day doing lab work in the station, repairing spacesuit gear, and engaging in other light activity.

The first thing we did after breakfast was perform a psychology experiment for the human-factors research group at NASA Johnson Space Center. The folks there are interested in knowing what kind of things would most brighten up the living environment in remote outposts like ours. So they sent us a set of seven posters. The backs of these ruggedly laminated sheets had served nobly and well as signboards during our Viking anniversary motor rally two days ago, but now it was time to use them with images displayed for their intended purposes. Six of the posters were of the travel agent variety, depicting verdant forests or tropical islands. The seventh showed a man standing above the clouds on top of a stark mountainpeak accompanied by an inspirational quote from Sir Edmund Hillary: "It is not the mountain we conquer, but ourselves." They wanted to know which posters we would like the best for our public space (i.e., the hab wardroom on the upper deck) and our private space (i.e., our bunks).

Since none of us spend any time in our bunks other than to sleep, the only decision that mattered was which picture to post in the wardroom.

We all agreed that all the lush travel posters would look absurd on the wardroom wall, where they would have to stand next to a window looking out on the bleak but strangely beautiful Haughton Crater. So despite the trite slogan from Hillary, the mountaintop scene won the day. Then we sat around the table and spun a Swiss army knife to determine the order in which each crew member would choose his or her poster. I came in third and selected one depicting "The Land of Myths: GREECE." I am a lover of Homer and in fact brought his works to the Arctic with me as my potluck contribution to the station library. I would very much like to visit his homeland some day. So now the poster is on the wall of my bunk, uplifting the spirits of my backpack and sleeping bag.

Around the time the art show was winding down, Christine stole a minute to sit down at the keyboard and pound out some more Mozart. She was a bit flustered, because the instrument only has five octaves while the music requires seven, but it sounded very good anyway. So I snatched one of the voice recorders we use for our mission reports and taped a sonata for transmission to Mission Support. We have a CD and tape player in the hab, but there is a stirring quality to live music that no recording can match. I think that a human mission to Mars would do well to include at least one person who is skilled with a musical instrument.

We transmitted the first concert broadcast from Flashline Station. I look forward to hearing the first one broadcast from Mars.

Cathrine and Charles spent much of the day in the lab, examining the geological samples we have acquired over the past week. Through an oversight, the geology lab lacked hydrochloric acid, a rock-testing staple. So Cathrine and Charles made do with vinegar. The soil samples fizzed but the rocks didn't, indicating calcium carbonate in the soil and dolomite (magnesium-calcium carbonate) in the rock. Martian ingenuity.

While standing on a high butte overlooking Von Braun from the north during the EVA yesterday, Cathrine had taken a series of fourteen overlapping photographs to provide the basis for making a panorama. During the early afternoon, while I puttered around working on our spacesuit backpacks, Brent used PhotoShop to stitch Cathrine's pictures all together to create a 360-degree view. It's pretty spectacular.

Life in an international crew continues to be an enlightening experience. Charles, our Frenchman, gets dozens of personal messages, which he is delighted to let the whole crew read. They are all from women. There is Dominique, Melissa, Kathy (the only one) and Kathy (the best), Lisa, Veronique, Françoise, and I think several others whose names escape me at the moment. Cathrine and Christine are in stitches over it, yet it must be said that there are a lot of men who would like to be in his position.

JULY 23, 2001

Our EVA today almost went perfectly.

The day was splendid, and the team consisting of Cathrine, Brent, Charles, and yours truly managed to get suited up and out the lock in just over an hour—a big advance over five days ago, when the same crew took three hours to get three people prepared. We then went over the ridge and into the crater, navigating accurately to reach our objective, a large erratic boulder at its predicted UTM location of 8368.9 N, 418.9 E.

We spent about an hour and a half at the boulder, examining it in every way. We measured it, we sampled it, we photographed it. If you can do it with a house-sized rock, we did it. Our goal was to figure out the history of the block, and Charles and Cathrine managed to work out what I think is a sound hypothesis. The boulder contained lots of shocked material, so it clearly came from the impact. But it was sitting on top of glacial debris, so it seems that it did not fly directly to its current location but to somewhere else, after which glaciers managed to relocate it. The block's composition is mostly dolomite—a limestone-like carbonate, which means that it did not come from very deep underground. It was covered with bacterial coatings of various colors, as well as different types of lichen, and we sampled them to allow for lab analysis to determine their nature.

Having met all our science objectives, we headed home, zipping across the crater in high spirits. I was traveling last in line and thinking how easy the EVA had been, when suddenly my ATV slowed, stopped, and sank into the mud lying just beneath the bone-dry covering on the crater floor.

I dismounted immediately to take weight off the vehicle, and tried to push it while running the engine, but we both just sank deeper. I radioed for assistance, and the others, who had been moving rapidly into the distance ahead, doubled back to help. I grabbed a rope off the back of my ATV, and moved around to tie it to the hitch at the front. I then threw the rope to Charles, who by this time was standing in slightly firmer mud about 10 feet in front of my vehicle. We tried to pull it out, but the two of us were not strong enough. I moved back along the right side of the ATV to try to reach the throttle so I could fire the engine while pushing. This was a mistake. The mud was much worse on that side of the ATV, and I sank down halfway between my knees and my hips. The whole affair was quite scary, because unlike the encounter with the mud a week ago, this time I could not detect any floor of permafrost under it. Also, the mud this time was more liquid, making it impossible to dig away from around my boots.

I grabbed the throttle. The ATV had stalled out, so I reached for the starter button, which is on the left side. The ATV would not start. It was still in gear. Somehow, I had to reach the left foot pedal, which was far down on the other side of the vehicle. But I was stuck almost hip-deep in the mud on the vehicle's right side and couldn't reach the left pedal. I got my midsection up over the rear of the ATV, and then pulled with all my strength to get a leg free and up onto the rear rack of the vehicle. Then, pushing with that leg, I got my other leg out. I was out of the mud and could now reach the gear shift, but the ATV was sinking deeper than ever.

By this time, Brent and Cathrine had made it back to where Charles was standing. I started the engine, revved it, and shifted into first. Then, with the three of them pulling on the rope, and me gunning the engine and pushing the ATV as best I could, we dragged the vehicle out.

The remainder of the trip back was uneventful, and quite pleasant except for the aftereffects of a massive adrenaline overload. We stopped at Trinity Lake to take water and soil samples for Christine's analysis, and then returned to the station.

JULY 24, 2001

Our plan for today had been for a major motorized EVA to the northeast, but when we awoke it was raining and blowing hard, so our plans changed. Today we would work inside.

In a way, the bad weather this morning was a blessing, because we had a lot of inside work to catch up on. The commitment of most of our manpower to EVAs over the past week had created a large backlog of unanalyzed and uncataloged samples. We were also behind on our reports. In addition, much of our equipment had begun to deteriorate, and we were running low on functional spares. So an inside day was really in order.

For the past two weeks, we have been trying unsuccessfully to do a voice interaction with the crowds coming to our soon-to-be-deployed Mars Desert Research Station currently on exhibit at the Kennedy Space Center Visitors Complex. One day our communication system would be down, the next day we would be fine but they would be shut down because of a hurricane. Yesterday all the technology was in place, but we failed to respond because I was stuck in the mud of Haughton Crater at the time. Well, today it worked. I got their questions as simple text, taped a recorded answer, transferred it into my computer, and hit the send button for transmittal just before the deadline of 5:45 P.M. eastern time. Apparently, the two-week series of glitches had created a sense of tension at the other end. When my answers finally played from the auditorium loudspeaker, there reportedly was applause.

Some deliveries arrived today. Among the most important were Incinolet liners and wax paper. The Incinolet is our toilet. It uses an electrical heater to incinerate human waste, thus reducing it to sterile dust. The system could be good for a Mars mission, because it requires no water (a standard toilet uses 2.5 gallons per flush, about the total of what each crew member uses here in a day). It does require power (about 3.5 kilowatts), but if the Mars mission has a nuclear reactor power supply, as it almost certainly will, that won't be a problem. The only mass the toilet requires is a kind of folded wax paper liner, which you insert into the stainless steel bowl. You do your business on that, then hit a lever and the bowl below it opens up like a set of bomb-bay doors. Then bombs away, the stuff is dropped into the incinerator chamber below. The purpose of the folded wax paper liner is to keep the steel bowl clean.

Around the middle of the second crew's rotation, we suddenly realized that we were about to run out of Incinolet liners. The units were new to us, and we hadn't realized that the box of liners sent with it represented only a small sample. I suggested that we substitute paper towels, but Steve, expressing fears that such unorthodox practice could result in someone ending up with a hot tush, insisted that we revert to using a camp toilet (i.e., poop into plastic bags) instead. I rejected this course, because it would stink up the station. Fortunately, we located a roll—exactly one roll—of wax paper in Resolute Bay, which served as an interim solution. As of yesterday, however, that roll was close to its end. Gratefully, today we received two replacement boxes of Incinolet liners, plus four rolls of wax paper forwarded to us by a Mars Society member in Edmonton. Many thanks to Mission Support for arranging this rescue operation in time.

(As an added plus, the timely arrival of new Incinolet liners provided Cathrine with the translucent material she needed to start drawing her map.)

The other item that arrived was a treadmill exercise machine. Cathrine and I assembled it on the lower deck after dinner. I was afraid we would have no place to put it; with all the tools, the electrical bench, Steve's communication gear, the glove box, EVA equipment, the geology, biology, and chemistry labs, and other things, it is getting pretty crowded down there. But fortunately, the Canadian Mars Society member who selected the treadmill (Darlene Lim) had chosen wisely: it is the fold-up kind that can be stowed away when not in use.

I can't understand why everybody likes treadmills. You can get just as much exercise running in place without a machine. But everyone does, and I'm no exception. We all got a kick out of using it. If a human Mars mission does decide to use exercise equipment, they'll certainly want to choose something lighter. This thing weighs a ton. But at least its sound drowns out the drone of Charles's rock saw.

JULY 25, 2001

Our numbers were strengthened today by a new crew member, Lieutenant Colonel John Blitch. John is here to replace Steve Braham, whose continued participation in the simulation has been made impossible by his burden of responsibilities for the Haughton Mars Project (HMP) communication system. John, a U.S. Army robotics expert, was scheduled for participation in the fourth crew rotation, which begins Saturday. So bringing him into the station early was a simple move.

It is true that on Mars it will not be possible to switch out crew members as needed. Then again, crew members on Mars will not have other jobs that make their replacement necessary.

In any case, John's appearance was a welcome one—it not only added to our manpower, he brought robots along as well. The three machines that accompanied him looked like futuristic black toy tanks with characteristic dimensions and mass varying from that of a coconut to that of a watermelon. He demonstrated these devices to the crew, who were all pleased by the capabilities they offered, but disturbed by the news that none of the little fellows had names. We quickly remedied this glaring defect, dubbing the creatures from smallest to largest: Stumpy, Jan, and Titan.

These three robots were all tethered devices, controlled by wire. John's wireless robots will be brought to the station tomorrow. The question is, What practical use might such machines (whether wired or wireless) be to human Mars explorers?

On the basis of our simulation activity here on Devon, it has become clear that once you have human explorers on Mars, it makes no sense to send robots anywhere the crew can go. Indeed, human explorers so vastly outclass robots as investigatory agents that it is simply a waste of time to send a robot to sites accessible by humans. This is true even though not every member of the crew can go EVA every day. Crew time inside the hab is almost as valuable as crew time outside, and to waste it by assigning a hab-based crew member to spend all day trying to get a robot to accomplish what an EVA astronaut could do in two minutes is just silly.

But there are places where an astronaut cannot go. One such place is under the station. The commander of a Mars craft might be very interested in the condition of the tanks and fuel lines under his vehicle. If his hab stood as close to the ground as Flashline Station does (which is strongly desired for EVA purposes), it would be next to impossible to get a suited astronaut under it to carry out an inspection. Similarly, I would like to know the condition of the structural supports beneath Flashline Station, but it's impossible to get an EVA team under it. So send in the robots.

The robots were carried outside by Charles, Cathrine, and me before we set off on our reconnaissance EVA to the north. I hooked up Stumpy to his outside control cable, leaving the little guy enslaved to the combined will of Brent and John, his masters inside the hab. Then we took off toward the Von Braun Planitia, with Inuit Ranger Joe Amarualik riding shotgun to protect us from polar bears (two of which were sighted by the HMP helicopter yesterday).

We spent the day exploring a number of fascinating canyon systems to the north of the Planitia. When we got home, we heard that Stumpy had had a great day too. Directed by his intrepid commanders Brent, John, and Christine, the robot had inspected the underbelly of the hab and imaged its supporting structure. Fortunately, its data on that matter were reassuring. Stumpy was so successful at his underworld inspection job that Brent, as station capcom, requested that we wait outside while Stumpy examined the underside of our ATVs. Stumpy got under my ATV well enough, then using his headlights for illumination, slowly and carefully imaged every feature of its substructure. This process annoyed me more than a little, since I had been in a spacesuit for six hours and really wanted to make use of the hab's facilities. So I politely suggested to Brent that Stumpy could do his work just as well without a live audience, and we entered the lock.

Later, we held a debriefing and worked out our plans for the next day. The plan is this: Tomorrow we will take robots with us into the field. Specifically, we will return to the rim of the northern canyons with Titan and a comparable wireless model named Talon, and send them in. We will see what robots and humans acting in a combined operation can do.

JULY 26, 2001

I looked out the porthole window this morning and saw nothing but white fog.

We were socked in. This forced a change of plan. We had intended to perform an EVA to deploy telerobots into the canyon that Charles, Cathrine, and I had scouted yesterday 7 kilometers to the north. But the fog killed that idea. Two polar bears had been spotted in that area yesterday by the Haughton Mars Project helicopter, and the thick white fog would give them all the cover they would need to take us by surprise.

Plotting a new route to an alternative canyon to the east took some time, then we had some electrical problems with the Incinolet, followed by some with the rovers, after which there was the usual pre-EVA fidgeting about looking for necessary items. As a result, by the time we were ready to suit up, it was lunchtime, so we had lunch instead. The bottom line was that we did not get out the airlock until 2 P.M.. This actually turned out to be for the best, because by that time the fog had lifted. We could have reverted to our original northern objective at that point, but we stuck with our new plan.

The objective was now to deploy a telerobot into Devo Rock

Canyon. The canyon is only about 3 kilometers from the Flashline Station, making it a good first target for field telerobotic work. The EVA team would be composed of John, Brent, Cathrine, and me, with Charles serving as station capcom. Our two telerobots would be Titan, the largest of the tethered units, and Solon, an even larger radio-controlled system.

We loaded the robots onto the front of Brent and John's ATVs and set off. We had to enter the crater on our way there, and as a result had a minor problem when John's ATV got stuck in the mud. But we got him out quickly and experienced no further difficulties until we hit the high ground and had to climb up a boulder-covered slope, where my ATV got hung up on some rocks. We pulled it out, and pressed on. By the time we made it to the canyon, it was positively sunny, and those of us who had dressed warm were sweating profusely under our suits.

The canyon looked steep and deep to me, but John was disappointed with it. The problem was the slope was only about 50 degrees. A human could climb that slope, John said, so it really would not do as a telerobot deployment site. He pointed across the canyon to a sheer cliff on the far side. That's where we should go, he said. Cathrine agreed with him, observing that the geology over there would be much more interesting. But it was already pushing 3:30, and Brent and I were both skeptical of the ease of access to the opposite cliff, which could only be approached by doubling back quite a ways to reach gentle slopes down to the river and then up again on the other side. For this reason I rejected the suggested redeployment. This displeased John, but we scouted around and found a site on our side of the canyon with a 70-degree slope, and he was satisfied.

The first robot to be deployed was Titan, the giant of yesterday's tethered group. I positioned him at the canyon's edge with his cable attached, while John set up the control station on an ATV and showed Brent how to operate it. Instead of a computer monitor, for the sake of lightness we used a three-inch camcorder display as a viewscreen to allow the operator to see what Titan was seeing. We did some function checks, then Brent initiated the forward roll command, and the small turretless black tank went over the rim.

Titan dragged his cable down behind him, but the wire frequently got hung up on sharp stones. Whenever this happened, I, standing near the edge, would shake the line free, allowing the robot to proceed. That kind of help is not cheating: Titan is a telerobot and is meant to be used by human explorers. Actual astronauts on Mars could provide their telerobots with the same assistance.

With some help from his friends, Titan made it about 100 feet down the slope. He had to stop there, because we ran out of cable. But overall it was an impressive performance.

Then it was Solon's turn. Though Solon is controlled by wireless radio, we attached 200 feet of parachute cord to him so we could pull him out in the event he got stuck. He is called Solon because he wisely projects a laser grid in front of him as he goes along, allowing him to measure everything he sees and to better deal with terrain obstacles. Unfortunately, some of Solon's settings had been set incorrectly, and he moved so slowly that he didn't get far before we decided to pack him back up. We found the problem when we returned to the hab, so he should be okay in the future.

We didn't accumulate much scientific data from our telerobots today, but that was not the point. Telerobots are not replacements for people; they are tools for people. In today's exercise, we demonstrated that a team of human explorers operating under EVA constraints could take a pair of handy robotic tools into the field and deploy them for use at a remote control station in a realistic operating environment. That's what I call a good day.

JULY 27, 2001

The weather is getting worse. The forecast for today was rain, sleet, and fog, with snow possible tomorrow. Most of my crew needs to be on the 737 that will leave Resolute for Yellowknife Saturday afternoon—the next flight out is Wednesday. So all plans for a telerobot deployment EVA to the cliff across Devo Rock canyon were canceled. Instead our program was to do our work today in the station with most of our bags packed, holding ourselves in readiness for rapid pullout to Resolute on the next Twin Otter flight, as it might prove impossible to get one on Saturday.

In fact, it proved impossible to get one today. So we ended up putting in a full day of work in the hab. Charles and Cathrine completed the logging of our rock samples. Then, while Charles wrote up their results as a kind of catalog, Cathrine continued her work of creating a combined terrain and geology map depicting the results of our survey expeditions to date.

Christine spent most of her day doing further chemical analysis of soil samples, Brent reviewed videotaped data from yesterday's EVA, and I worked on dealing with backlogged communications over logistics and other issues.

But John Blitch had a research program that needed the involvement of the entire crew. He wanted to find out what kind of people make the best rover operators, and to see if such abilities could be tested for in advance. Moreover, he wanted to do the testing on a group of people who are representative of the skill and character types one might actually send to Mars, and do it not when they are fresh as daisies, but when they have been fatigued a bit by extended fieldwork, just as they really would be on a Mars expedition. For John, then, our crew represented ideal test subjects. As this kind of research is quite important, we all gave him our full cooperation.

The tests started off with written exams in spatial abilities, memory, and pattern recognition. There were also problems that were kind of like tangrams on paper, in which you had to try to determine which groups of shapes could be put together to make a certain pattern, and others where you had to predict how various three-dimensional shapes viewed from one angle would be seen from another. Ordinarily, I'm pretty good at these kinds of tests, but I have been getting only four or five hours of sleep each night for the past several weeks, and so I found them a bit taxing.

After the written tests, he took us one by one down to the EVA preparation room. There he had set up a rover control station for Stumpy the tethered telerobot, who was positioned under the habitat. We were asked to drive the rover to three of the six habitat legs and find and read six little signs proclaiming habitat structure malfunctions that John's base camp assistant Arnie, operating outside the hab, had placed on the backs of the legs and on the underside of the habitat floor. You controlled Stumpy with a joystick, and could also pan his camera eyes up and down and change the depth of focus. The images from Stumpy would be transmitted back over his cable and displayed on a camcorder viewscreen, thereby allowing you to drive the vehicle and search for and read the little signs. You had to do this micro-reconnaissance as fast as possible, with John timing you to assess your performance.

I found it relatively easy to navigate the vehicle—that is, figure out where under the hab Stumpy was, and then set an accurate course toward the next major objective. But the ground under the hab is covered with irregular rocks of various sizes. While it was easy to tell which of the largest rocks represented impassable barriers that had to be steered around, I found it impossible to determine in advance which of the much more numerous small and medium-sized rocks Stumpy could climb over and which would hang him up. So Stumpy kept getting stuck. Sometimes we could get him unstuck simply by going into reverse, other times we had to jerk him free by pulling on his tether. So the recon was a tedious process. I took seventeen minutes to do it; other people's times ranged from around ten minutes to close to an hour. If the signs had been in places accessible by humans, any of our crew members on EVA could have done it in less than a minute. This exercise reinforced my view that the primary value of telerovers is not general exploration but exploration directed toward critical sites not accessible by human astronauts.

Remarkably, when John compared our written quiz scores with the results of the rover driving test, he found nearly a direct correlation.

Dinner tonight was prepared by Cathrine, who copied my recipe from yesterday of spaghetti with (canned) tuna fish mixed in marinara sauce. Imitation is the sincerest form of flattery. I was quite gratified.

This was our last night together as Flashline Station crew, so we held a going-away party. We played Martian Chess in teams (with no consultation allowed between partners), while the boom box blasted away the tunes of Fleetwood Mac well into the A.M. hours.

We need to make it to Resolute tomorrow. Will a Twin Otter show up in time?

JULY 28, 2001

When we awoke this morning there was no fog, but a low ceiling, and the word from camp was that Twin Otter flights could not be scheduled under these conditions. But around 9 A.M. we got a radio call from HMP camp manager John Schutt informing us that First Air was going to give it a try, and that we should be at the landing strip with all our stuff in forty-five minutes. Knowing that any lucky break in the weather would be brief, we were already packed and ready to go.

We started passing duffel bags and packs down from the upper deck. As we did so, Pascal came zipping up out of camp on an ATV, accompanied by an Inuk lad driving another ATV pulling a small trailer. As the trailer was loaded, I gave Pascal a quick tour of the hab, briefing him on its current status, showing him where various key items had been stowed, and clueing him on the minor quirks that had developed with various pieces of equipment.

Then we were off to the airstrip. I had Christine riding on the back of my ATV, and we made the trip down the ridge to the Lowell Canal quickly. The stream looked low, but I did not want to be caught in mud. So I increased speed to try to quickly cross at a particularly shallow-looking spot, and instantly submerged the ATV in two feet of ice-cold water. The engine stalled out, and for a moment it seemed as if Devon Island was going to get in one last shot at complicating our plans. Fortunately I was able to restart the engine, and we got to camp with refreshed feet but little else worse for wear.

The whole camp turned out to see us off.

ROTATIONS 4, 5, AND 6

There remained three more crew rotations scheduled for Flashline 2001 after I left the Arctic. Rotation 4 consisted of Pascal, Carol Stoker, Larry Lemke, John Blitch, Peter Smith, and Charles Cockell. This rotation lasted only five days, but went well. The crew used John Blitch's rovers to scout a site in advance, and then followed that up with in-person exploration of the same site. They thus replicated the model that many in NASA believe to be the necessary standard operating procedure for extraterrestrial exploration: first send robots, then humans.

Interestingly, however, the results of this exercise demonstrated a negative conclusion. There was no value to sending the robots to the site in advance. The robots missed much essential information about the site—for example, the presence of lichens—and all the valid information they did return was readily apparent to the crew within the first minute of their arrival on the scene. Thus the crew had spent an entire day running robots around a site with no added scientific return whatsoever.

Far from being a waste of time, however, such activity conducted in sim on Earth is extremely valuable, since it will help us avoid wasting time and money developing systems for use in similar ineffective tactics on Mars. Basically, it showed that surface robots are not appropriate systems for advance scouting for humans. If you want to do reconnaissance for humans, you need highly mobile systems like orbiters and aircraft that can help plan routes and identify objectives for the astronaut explorers who will really bring in the scientific bacon.

While Crew Four was a solid success, marred only by Pascal's frequent refusal to interact with Mission Support, Crew Five was not. The problem was that, in contrast to the very strong composition of Crew Four, in Crew Five there was no one with sufficient professional stature and force of personality to keep Pascal on track. According to some of the Mars Society volunteers who served in Crew Five, Pascal started freecycling (which means getting up and going to sleep at random hours uncoordinated with the rest of the crew) and going off to HMP camp to participate in activities there. This left the crew adrift. Some useful testing was done of a two-person ATV designed by Purdue University. But virtually no reports were filed and comparatively little science was done. Caught up in the chaos of the field season's close, Crew Six was a shambles and was mercifully canceled after only a few days of activity.

Nevertheless, on the whole, the Flashline Station summer 2001 field season on Devon Island had been a grand triumph. For the better part of half a century, people had talked about conducting Mars mission simulations in the Arctic, and we had finally done it. In doing so, we had demonstrated in dramatic fashion the overwhelming value of human explorers on Mars. Under Mars mission simulation conditions, we had performed broad-ranging field surveys, crossing or climbing over very tough terrain. We had found stromatolites and cyanobacteria, and we had deployed seismic arrays and telerobots. These were all essential activities for Mars exploration, and none of them would have been possible using strictly robotic means. We had shown that a Mars crew could operate without cost to morale with a water ration less than half that assumed by NASA. This was a powerful result that could have real impact in reducing the mass, and thus the price, of a human Mars mission. We had learned an immense amount that would help us refine the simulation for the next year. We had reached tens of millions of people through the media that came up to cover us. In short, we had done it all.

On the last weekend in August, most of those involved journeyed to the Mars Society Convention at Stanford and took a bow. Great plans were laid for a vastly expanded simulation program to be initiated at the Mars Desert Research Station during the coming winter.

Two weeks later, the World Trade Center was destroyed.

9. INTO THE DESERT

The year was 1846. Even as Sir John Franklin was coasting along Devon Island on his final disastrous voyage, far to the south events of epic proportions were unfolding.

Americans in California had raised the Bear Flag of revolution against the corrupt and ineffective Mexican government and the Lone Star Republic of Texas had declared for union with the United States, leading to war between the U.S. and Mexico. War also loomed between the United States and the British Empire over the Pacific Northwest, as American and British diplomats exchanged incompatible border claims based respectively on the 54th and 42nd parallels.

The fate of the continent would not be left in the hands of governments, however. Even as the U.S. Army under General Steven Watts Kearney was marching on Santa Fe, a much larger nonuniformed force was moving west to seize a greater prize. They were called the emigrants. In wagon trains emblazoned with the bold cry of "Oregon or Bust," in 1846 they hit the trail in hundreds of thousands to attempt to realize Manifest Destiny's call for a continental nation based on liberty. Many, such as the Donner Party, which set forth that year, did not make it, but most did, and in doing so they changed the demography of the continent in a way that no diplomatic negotiation could ever reverse.

The noted Boston patrician historian Francis Parkman was journeying through the West during that epic year, seeking material on American Indians to incorporate as background for his works on the previous century's French and Indian Wars. A genuine snob, Parkman was not at all interested in the incredible historical events that were unfolding all around him, and tried as best he could to avoid the emigrant wagon trains.[30] This proved impossible, however, and Parkman's involuntary encounters with the westward-bound emigrants ultimately led to the writing of his classic, *The Oregon Trail*.[31]

If you read this book, it is very clear that Parkman did not understand the emigrants. Some (the New England men, of course) were a cut above the others, but all were fundamentally irrational. Why seek new land in Oregon—and the term encompassed the whole Pacific Northwest, including the modern states of Washington, Oregon, Idaho, and Montana, and northern California—when there was plenty to be had in the much nearer unclaimed vastnesses of Ohio, Indiana, Illinois, Wisconsin, Iowa, and Missouri? To Parkman, it made no sense.

If the Oregon emigrants were irrational, however, there was another group that Parkman found utterly insane. These were the "fanatics," adherents of a new religion proclaimed by a recently deceased prophet named Joseph Smith. While ignoring nearer and safer alternatives, the emigrants at least were heading toward virgin lands that were known to be of high agricultural quality, and their Manifest Destiny ideals, while not shared by Parkman, were at least comprehensible within the context of American culture. The fanatics, on the other hand, were guided by alien and fantastical religious concepts, and instead of setting their course for balmy Oregon, were madly heading into the desert toward a place they called Zion. The emigrants might be foolish, Parkman thought, but the fanatics were crazy.

Parkman's "fanatics" were the people that are today generally called the Mormons, or as they call themselves, the Latter-day Saints, and their Zion was the place that is today known as Utah. The Mormons are of great interest to those who would consider the dynamics of human colonization efforts, because unlike certain others, such as the Oregon emigrants, their idealism could not be hid behind a smokescreen of nominal practicality. There was no valid or even distantly plausible economic justification for the Mormon settlement of desolate Utah. One might argue that they went to escape persecution, but since the Mormons were ethnically and culturally identical to the mainstream Protestants who were persecuting them, a much simpler means to escape persecution would have been to just return to the religions of their birth. So strictly speaking, they did not go to Utah to escape persecution, *but in order to be what they wanted to be,* and to live in a society constructed in accord with their own ideals.

In other words, it was about fundamental freedom—the right to cut one's own path and make one's own world. For such freedom there have always been, and I believe there always will be, people wishing to sacrifice wealth, comfort, and safety. This is the motive that two hundred years before the Mormons, sent the Pilgrims to build their City on a Hill on the bleak shores of New England. It is also the motive that will, I believe, two hundred years after the Mormons' migration, inspire human colonists to try their courage and grit on Mars.

Utah's pioneer past is Mars's future.

There could thus be no better place to establish a Mars Desert Research Station.

DEPLOYING THE MDRS

We had planned the deployment of the MDRS for the fall of 2001, but the terrorist attack of September 11 sent things seriously out of kilter. It wasn't so much the direct effects of the attack itself, although several of our New York City members came very close to getting killed. Paul Contoursi, the head of our New York Chapter and East Coast fundraising task force, worked on the 97th floor of the World Trade Center, and was only saved because he stopped on his way to work to vote. Emerging from the subway, he saw the first 767 fly into his office area and explode. John Blitch was part of the postattack rescue effort, and several of his robots, including Solon, who had been with us on Devon Island six weeks earlier, were lost in attempts to locate survivors in the wreckage. As traumatic as these events were, however, it was the attack's economic and political fallout that produced the larger impact on our program.

Almost immediately after September 11, the money dried up. As late as the summer of 2001, there had still been a lot of youngsters hanging around Silicon Valley owning dot-coms with nominal values of tens of millions of dollars. These kids had grown up watching *Star Trek,* and were a major financial resource for us. Now they were broke.

Then there was the new political reality. The War on Terrorism threatened to sweep away space exploration as irrelevant to the needs of the day. To answer those who espoused refocusing the space program on purely military functions, I wrote a major op-ed in the aerospace industry weekly *Space News,*[32] in which I argued that an aggressively expanded space exploration program could play a decisive role in winning the war against Islamic fundamentalist terrorism by demonstrating the superiority of a civilization based on the use of human reason. This article was seized upon by a couple of Internet incendiaries to flame me as being an American chauvinist and "anti-Islamic bigot" and launch a factional disruption operation in the Mars Society under that banner. Dealing with this occupied a very trying month of my time.

In the end, however, the membership rejected the slander campaign. The Mars Society membership also pulled us through financially by responding with some $32,000 in small donations to a direct fundraising appeal. Together with a $25,000 donation from the Greenleaf machine tool company that came through in the late fall, this gave us the financial reserves we needed to deploy the MDRS.

So in December 2001, the components of the MDRS were belatedly shipped from their storage warehouse in Denver to the construction site in the desert about five miles northwest of Hanksville, Utah. Frank Schubert went out to lead the construction, assisted by a group of Mars Society volunteers mobilized by University of Michigan biochemistry student Anna Paulson. A Utah native herself, Anna had been a leader of the successful University of Michigan student project to build an analog Mars pressurized rover. In the fall of 2001, she took a semester off from school to help us get the MDRS program going. Attentive to the needs of volunteers, she proved a highly capable right hand to Frank during the MDRS construction phase.

As at KSC, the combination of the MDRS's lightweight components and the ready availability of construction equipment made erection of the station in the Utah desert much easier than on Devon Island. With the help of Anna's volunteers, Frank was able to assemble the primary MDRS structure in less than a week. By the beginning of Christmas week, the interior had been partly built out, and in a very limited sense the station was ready for occupation. Frank then went home for the holiday, bringing construction to a halt, but Anna made productive use of the week to lead the volunteers in a preliminary shakedown rotation.

Because of the crude nature of the MDRS at this time (there was not even a communication system), and the lack of scientific equipment, complete spacesuits, ATVs, and other necessary gear, Anna's shakedown was even more rudimentary than Carol's had been at Flashline at the close of the 2000 construction season, but her crew made a significant start of the exploration of the surrounding region and identified a number of important areas where the hab needed improvement.

Following New Year's, Anna and most of the volunteers had to return to school or work, leaving Frank with limited assistance. Working through January, sometimes alone, sometimes with the help of a local resident or a

Mars Society volunteer or two, he managed to complete the build-out of the interior of the station. Also during this period, I sent two of my Pioneer Astronautics engineers, Gary Snyder and K. Mark Caviezel, out to Utah to install the hab's satellite communication equipment. Highly adept at electronics, Gary and K. Mark had also created a unique UHF radio repeater system, and installed a solar recharged unit on an elevated position on the Skyline Rim miniature mountain range that ran in a north-south direction a few miles west of the hab. With the help of this system, crew members on EVA using handheld radios would be able to communicate with the hab despite terrain obstacles and over very long distances.

By late January, the station was ready for a second shakedown rotation. This one was led by Frank, and included Bo Maxwell, the head of the Mars Society U.K., and a diverse group of European Mars Society architects who wanted to gain firsthand experience with the MDRS to guide their thinking in the design of the soon-to-be-initiated Iceland station (EuroMARS) project. Because of limited four-day duration, lack of ATVs and lab equipment, and the absence of field scientists in the crew, the simulation activities of this shakedown were also quite limited. But they confirmed that the hab was ready for full-time occupation, got the habitability design insights they were seeking, and did some exploration as well. In addition, they certified the suits for desert use and identified a key bug in the satellite communication system that required correction.

Over the last weekend in January, our electronics wizard, Gary Snyder, went out to Utah to fix this problem. That job being successfully completed, the MDRS was finally ready for operation.

In November, we had put out a call for volunteers for crew slots, and as before with Flashline, received hundreds of applications from around the world. A committee had downselected these to a more limited pool, and from this group I selected six qualified six-person crews for the MDRS spring 2002 field season. Each rotation would do a two-week tour, and then train the following crew in station operations during switch-over. MDRS Crew One was scheduled to begin February 7. For its first week, I would be in command.

LETTERS FROM MARS BASE UTAH

As I had done during the previous summer in the Arctic, I documented the operations of the MDRS during my stay with a series of dispatches. These were posted daily on the Mars Society website (www.marssociety.org) and on MSNBC.com under the title "Letters from Mars Base Utah." The following selections give a good account of how things went.

FEB 7, 2002

After months of delays, the Mars Desert Research Station finally went operational today. A lot of things are still balky, the satellite communication system is behaving erratically, much of the internal network doesn't work, and there is a problem with one of the water pumps. But we have a completed and fully provisioned station, a fairly well-equipped lab, a good power system, five functioning spacesuit simulators, three good ATVs, sufficient satellite and local UHF com capabilities to function, and a highly qualified crew that is willing to do what it takes to push through. So today we began. . . .

The first crew, coming from various locations, met one another for the first time in Hanksville this morning. We then drove out to the hab together. Our team includes Steve McDaniel and Troy Wegman, both biologists. Steve is a Ph.D.-turned-attorney who works with the Texas Technology Litigators patent law firm. Troy does microscopy for the Mayo Clinic. There are also two women: Jennifer Heldmann, a planetary geology Ph.D. student from the University of Colorado, and Heather Chluda, an aerospace engineer who works on the Space Shuttle program at Boeing-Rocketdyne. The crew is rounded out by Frank Schubert, the Project Manager, who works as an architect, and me, an astronautical engineer. I'm in command, but only for a week. After that I will be rotated out and replaced by Tony Muscatello, a chemist who leads Mars Society Mission Support. Frank will also leave after a week to be replaced by Andrew de Wet, a geologist from Franklin and

Marshall University. Everyone else will stay for the full two weeks, after which another two-week-long volunteer crew will take their places.

We reached the station around mid-morning and worked together as a team hauling in the lab equipment and the provisions for the season, and following that, cleaning the place up. Then, while Steve and Troy set up the biology lab, Frank fixed various things, Heather and Jennifer programmed our mobile weather station, and I labored, with only partial success, to get the Starband satellite dish to work. I can see why these things aren't very popular. Believe me, if you have a DSL line, a cable modem, or a copper telephone line for that matter, don't get a satellite dish. These gadgets are finicky. Sometimes they work fast, sometimes very slow, sometimes they lose link in the middle of a transmission and you have to start all over again. But then again, the communication links from Mars won't always be so great. We'll fix it if we can, live with it if we have to.

We start field operations tomorrow.

FEB 8, 2002

We initiated EVA exploration operations today. The team was all first-timers—Heather, Jennifer, and Troy. They did a great job, and filed an excellent report.

Back at the hab, however, the day was hardly uneventful. We had a windstorm. While our weather station was unfortunately not yet operational, a conservative estimate was that it was blowing at least 60 knots. Part of the dome of the hab almost broke free, and the greenhouse tried to take off for Kansas, and Frank, Steve, and I had to break sim to deal with it. The aim of the roof-mounted satellite dish was also disrupted, causing us to lose Internet communication capability until evening. The most violent part of the storm was fairly brief, so that after the chaos subsided we resumed contact with the EVA team using our local repeaters. Frequently, however, the background noise caused by the wind blowing around their helmets made them difficult to understand and made it hard for them to hear us. When we would get their reports, we would repeat the essence of it, twice, and then ask, "Is that correct? Please respond affirmative, affirmative, affirmative or negative, negative, negative." That's what it took to distinguish between yes and no.

There are sometimes very high winds on Mars. Because the atmosphere is only 1 percent as dense as that of the Earth, however, a 100-mile-an-hour gale on Mars only packs as much force as a 10-mile-an-hour breeze on Earth. So astronauts won't have to deal with flying greenhouses. But the storm will still make plenty of noise. Today was an interesting test.

FEB 9, 2002

At our morning meeting, I laid out our plan for the next several days: a series of long-range motorized reconnaissance EVAs to give us a broad familiarity with the area and identify key sites for further in-depth study. One of the crew members asked if it might not be more methodical to start at the hab and slowly spiral out, studying one site after another in turn. My response was no: When you explore a house you don't walk in the front door and then stop and spend several hours examining the contents of the foyer with a microscope. No, you give the place the once-over first. It is the same with field exploration. Before you invest a lot of time in focused study of particular sites, you conduct a general survey. This gives you the overview you need to assign your priorities.

The EVA team was composed of Steve McDaniel, Jennifer Heldmann, Heather Chluda, and me. With four people going EVA, it took a while to get everyone suited up, so we were not out the lock until a little before noon. We took about twenty minutes to set up a weather station, and then headed north on our ATVs.

The weather was perfect. We set out heading north, and after traveling about 2.5 kilometers came across a rather impressive outcrop of sedimentary rocks. We decided to check it out. Jennifer, our geologist, and Steve, our biologist, collected all types of samples of rocks and possible cyanobacteria. I searched the place for fossils but didn't find much. This was a disappointment. The banded Mesozoic sediments included both terrestrial and marine materials, and wave ripples in the sandstone were clearly visible. By rights, the formation should have been full of fossils. It wasn't.

We continued north another 2.5 kilometers and came to a hill too steep for the ATVs. I decided to climb it, though, to get the view of the region to the west. We hiked up, and were rewarded not merely with an impressive view, but with the sight of a fair-sized canyon and a passable ATV route to get there.

So to the canyon we went. This was a wonderful place, with a steep little gorge that exposed millions of years of banded sediments to easy view. I climbed around the rim and had a Eureka moment when I found some bits of petrified wood. These were made irrelevant within minutes by Heather, who found a small mountain made of the stuff—in several varieties no less. But then I found something that really made my day—a bone of stone. It's the size of a coffee mug, and the indentation for the joint is clearly visible. The material I found it in was Jurassic, so my guess is that it's a dinosaur.

After the canyon, we continued further north, eventually coming to a huge cliff, with a 500-foot sheer drop past several epochs of exposed geologic history. The view was spectacular. Heather suggested we rappel down. That's the sort of thing she goes in for. Fortunately, however, no rope was available, and we all returned to the hab alive, having covered 19 kilometers in a day. . . .

FEB 10, 2002

The crew have been working so hard over the past several days that only one member has had time for a sponge bath, and it has started to get to people. This being Sunday, I decided to set aside some time this morning before EVA to give everyone time to wash. Unfortunately, we discovered that our water reserve tank was empty (we are still on a once-through water system—our recycler won't become operational until our greenhouse comes on line in March), leaving us with only 11 gallons left in the hab. I contacted Mission Support to reach our support person in Hanksville to come out and fill the reserve, but as there was no telling when this might occur, the sponge baths had to be canceled, and we went to the paper plates to cut the need for washing water to a minimum.

The upside of this was that it saved time. So we planned an ambitious EVA. The mission was to penetrate the ridge line of steep hills that runs north-south just west of the hab to be able to explore the large region of uninhabited land that lies between this local ridge and the even higher Skyline Rim that also runs north-south a further 3 kilometers west. The EVA team consisted of Troy Wegman (a biologist), Jennifer Heldmann (a geologist), and Heather Chluda (an aerospace engineer), with Heather in command. Their instructions were not only to try to find or force a pass into the region between the two ridges, but also to map the route with a series of UTM-gridded waypoints with verbal descriptions, photographs, and where appropriate, samples assigned to each. The idea is to create a guidebook to the area for the crew rotations that will follow us, so that anyone looking at our documentation would be able to know the character of the terrain around dozens of waypoints throughout the region.

The team left the hab a bit after noon and stayed out for almost six hours. While they were away, I acted as hab capcom and worked at improving our satellite Internet connection, with some success.

While I was thus engaged, Steve McDaniel, the other member of the crew who stayed in the hab, conducted lab analysis of the biological samples collected during yesterday's EVA. He imaged the samples at magnifications as high as 1000 times. The samples proved to be sublithic bacteria—exactly the type of organisms that some researchers believe could conceivably exist on Mars. . . .

Communication with the EVA team stopped after 3:46 P.M. This did not worry me excessively. There is rough topography around here that can cause radio cutoffs. When 5:30 rolled around and it began to darken, however, I became concerned. We are close to New Moon and there is no light pollution here, so when it gets dark it gets really dark. GPS could provide the crew the direction back toward the hab, but if they were caught in total darkness in rough terrain they would have great difficulty proceeding. Fortunately, at 5:50 Heather checked in, and they made it back—just barely—by nightfall.

When they came through the lock, they seemed both exhausted and exhilarated. It was obvious that it had been a great EVA. They had bags of fossil mollusks (lower Cretaceous oysters!) and other samples, and reams of data. Best of all, they had found a passage through the ridge. It's a rough trail, but well worth traveling. We've named it the Chluda Pass.

FEB 11, 2002

The crew was up till after 2:30 last night writing reports, and so were slow to awake, and it was 11 A.M. before we were ready to start suiting up. This made it unlikely that we would achieve our planned objective of reaching the Factory Butte area, some 20 kilometers away, by ATV through difficult terrain. But we decided to try to push as far in that direction as we could. Frank Schubert had returned to the hab the night before, and he showed me a pass through the local ridge that was discovered by one of the shakedown crews. As this route was easier than the Chluda Pass found by the EVA yesterday, our rate of progress would be improved, at least for the first part of the trip. The team would be Heather, Jen, Troy, and I—a four-person team being best for a difficult EVA as it provides extra muscle power to lift the ATVs should they get stuck. Four people also take longer to suit up than three, however, and it was not until 12:15 that we were out the lock.

We crossed the local ridge via the Schubert Pass and entered the large region we call Mid-Ridge Planitia, which lies just north of the lower Blue Hills. Part of this plain is scrub grass, but much of it is unvegetated Cretaceous marine sediments. Here we found huge fields of fossil oysters similar to the ones found yesterday. These we have now identified as *Pycnodonte newberyii*, a species that inhabited the Mancos

Sea, which covered much of Utah 85 million years ago. Interestingly, while we found oyster shells by the millions, no other species were readily in evidence.

We proceeded generally north across the Mid Ridge Planitia. Much of the travel was easy, but periodically the plain was cut by sharp little flash-flood channels too steep to drive down and then up in an ATV while wearing a spacesuit. So we had to dismount and push them across by hand. This slowed our progress, as did the necessity of making waypoints every kilometer or so. This latter process is accomplished by stopping and taking a GPS reading, several photographs, a radio link check, and a very brief geological examination to characterize the site. All of this is noted down on a chart that the EVA team carries and then included in our trip report. Our intention is to compile all this waypoint information into a kind of guidebook to the area for future crews.

As we went north, the landscape became increasingly barren, and spectacularly scenic in its bizarre desolation. The elevation changed periodically, and gray Cretaceous peaklets alternated with red Jurassic formations. Finally, we entered a region that has to qualify as a geologist's paradise: a chaotic assemblage of miniature canyons and outcrops of every description. We call this alien region the Barsoom outcrops, after the Martian world depicted in the romantic adventures of Edgar Rice Burroughs.

While rich is geology, the Barsoom outcrops are also rich in obstacles, which would make the remaining 2 kilometers to the Coal Mine Wash path to Factory Butte slow going. As it was already 3:30 P.M. when we entered the outcrops, it was clear that we could never make it to Factory Butte and get back to the hab before dark. So I decided to call a halt to the advance and have the team spend an hour exploring and sampling the Barsoom outcrops themselves. We did this, and then turned back, leaving Factory Butte for another day. Because we had already mapped out the outbound route, and were not taking waypoints, we were able do the return trip three times as fast as our outward trek, reaching the hab in early dusk around 5:30.

After dinner we wrote up our EVA reports and sent them to Mission Support, along with biology lab results developed by Steve and a hab engineering report from Frank. This was done by around 10:30 P.M., which is early relative to what we had managed previously. So we had a little meeting to discuss what to do the next day. The consensus was that it was important that we take some time to analyze and organize the large amount of samples and data we have assembled.

This settled, we had some time for some group R&R—i.e., a movie. Every member of the Mars Desert Research Station crew brings a few books, tapes, CDs, and DVD movies to donate to the hab, so we have a nice little potluck entertainment library. The crew's choice this evening was *The Matrix,* a film I had never seen before. I understand that many people consider this to be a very profound work of art, with its deep deliberations about what is real and what is not, etc., but there are more holes in its plot than can be found in a family-sized box of Cheerios. So I decided to view it as a comedy and found it very enjoyable when considered in that way. The experience was heightened by the availability of a little bit of rum, which was distributed to celebrate Heather's twenty-sixth birthday.

The film was done at 12:30. I hit the sack and slept like a rock.

FEB 12, 2002

We did not go EVA today. Instead we spent the day doing scientific work and maintenance around the hab.

The decision to stay at home today was a good one. We got a great deal done that went far toward producing concrete results from the hard fieldwork the crew has put in for the past four days.

Jennifer took the time to classify and catalog the large geological collection we have gathered, while Troy organized the biological materials and extracted samples. Steve took selected samples and subjected them to chemical analysis for specific enzymes. I wrote reports, dealt with logistical issues, and worked at transferring more of our data to Mission Support. Our Starband communication system remains very cranky, and frequently cuts off in middle of transmissions, thereby making it necessary to resend the same data again and again to get it through. Probably the most productive work of the day was done by Heather, who organized our database of EVA waypoint locations and characteristics into a single spreadsheet that will serve as an invaluable guide to future operations.

Assisted by Steve and Heather, Frank transferred the weather station from its test location outside the hab to the top of the dome roof, where it is now collecting good data. (For safety reasons, this operation was done out of simulation.) Frank also got our flagpole up, and the red, green, and blue Martian tricolor now flies proudly over the Mars Desert Research Station.

Dust is proving to be a significant problem. Some of the digital cameras have experienced malfunctions due to dust, which, by interfering with the opening of their zoom lenses, has frequently prevented them from turning on. I have recommended to Mission Support that future crew members be advised to bring cameras with no external moving parts if at all possible. We also have initiated measures to prevent dust from being tracked into the hab. From now on, no EVA suits returning from the field or boots or shoes of any kind will be allowed farther into the hab than the EVA preparation room. As Mars is also dusty, similar measures will probably have to be implemented there.

Having the entire crew working together in the hab tends to bring compatibility issues to the forefront. The key points of friction tend to be food and music. The former is not too bad. If someone really doesn't like what is cooked for dinner (we each cook in rotation; the cook dictates the menu), he or she can abstain or make something else. But whatever music is played on our sound system is heard by everyone. My musical taste runs to Beethoven and Mozart, while Steve McDaniel and Frank Schubert prefer contemporary groups with names like the Spiced Crows and the Potted Owls. The rest of the crew falls somewhere in between. My resolution is give each crew member a chance to choose the next tape in succession. This is not fully satisfactory, since everyone still has to listen to stuff they despise. But all agree it is fair, and that which is fair can be endured.

FEB 13, 2002

We had visitors today: a film crew for the popular German Science TV program *Nano* and a photographer for the *Los Angeles Times.* They wanted a story and we gave it to them. We took them with us on our EVA to Lith Canyon. The EVA team consisted of Steve, Troy, Jen, and me. Frank also came along, operating out of sim to drive the film crews to the site, while Heather stayed behind to work further on her classification of our waypoint database.

So out we drove, with the pickup truck in the lead, its flatbed filled with cameramen focusing on us as we followed in single file on our ATVs. We traveled in this fashion, along the dirt road we now call the Lowell Expressway, about 4 kilometers to the north, after which the EVA team peeled away to head off road to the west and the canyon site. Since we had identified the location and its UTM coordinates during our scouting expedition to the area Feb. 9, we found the site without difficulty.

We then descended into the canyon, moving systematically along the walls. Jen concentrated on geological analysis, and Troy and Steve did sampling of suspected endolithic bacteria. I did photo documentation and searched for fossils.

The day was sunny, and unlike the previous sorties, where we occasionally walked but spent the large majority of our time on our ATVs, on this EVA we hoofed it for hours. With the awkward EVA gear, the strong sun, and our increasing load of rock samples, the trek down the canyon became somewhat tiring—so much so that the media cameramen got exhausted, and opting for the better part of valor, asked Frank to truck them back to the hab.

The loss was theirs. As we continued down the canyon, the scenery became increasingly weird and the geology more interesting. We also spotted tracks of antelope and mountain lion. After three hours we went through a narrow pass, the canyon then opened up into a vista of bizarre formations, and then the floor of the canyon dropped away to a new boulder-strewn bottom 30 feet below. When a flash flood sends water flowing down the canyon's bed, this place must be a little Niagara, so for lack of a better term, I call it a dry waterfall. I really did not want to climb down the fall to the boulder field, but Jen was excited about the geologic potential of what lay at the bottom, and without further ado, made a nimble descent. This left the rest of us little option but to follow. I did a radio check to make sure that we could still access the repeater link to the hab should we need help, and that verified, started the scramble down, entrusting my soul to the designers of our spacesuit's U.S. Army cold-weather boots. Bouldering on soft sedimentary rocks in these clunkers while wearing an EVA simulation suit is an interesting experience, but we all made it down okay.

I don't know if it was worth the risk, but it must be said that Jen's intuition of the geologic interest of the boulder field below the fall was correct. She collected a variety of rocks and minerals not seen by us here before, Steve discovered a fossilized bone, and I found a magnificent piece of petrified wood the size of a volleyball. Naturally I had to bring the thing back, which made the climb back up the dry waterfall and the return hike through the canyon even more memorable.

When we got back to the hab around five, the Germans were still there, ready to interview us. I made them wait a little, while we had a debriefing of the EVA crew. Overall the EVA was completely successful. We did, however, take a risk. It could be argued that the risk was small, and certainly taking risks is sometimes necessary if results are to be achieved. Going to Mars will require accepting all kinds of risks, including several big ones associated with major maneuvers like space launch and interplanetary travel and innumerable little ones comparable to our climb down the dry waterfall today. The point is not to avoid all risk, or even to minimize it. The point is to calculate all risk; to take risk, but do it with due deliberation.

We may have fallen short in that respect today. But no matter, we are here to learn.

FEB 14, 2002

Today was changeover day. Frank's replacement, geology professor Andrew De Wet of Franklin and Marshall University, joined us last might. My replacement, Tony Muscatello, the head of Mars Society Mission Support, showed up around 11 A.M. Frank and I spent about two hours briefing the new arrivals. Shortly before 1 P.M., we gathered the crew for a brief parting ceremony on the upper deck. It was a warm moment; in our week working together we had all become good friends. It was time to leave, however, and a few minutes later we were gone.

With the completion of its first week of field research, the Mars Desert Research Station can now be considered fully operational.

Led now by Tony Muscatello, Crew One continued for another week before concluding on February 21. We had initiated scientific research across a range of disciplines, and set a good pattern for all future rotations to follow. But perhaps the standout accomplishment of Crew One was the initiation of the Waypoint database. This was the work of Rocketdyne engineer Heather Chluda, and illustrates the value of having intelligent non-specialists on a mission (since Heather's actual professional specialty, Space Shuttle Main Engine design, was not applicable to MDRS operations). Not being a biologist or geologist, Heather could not contribute much to the field science directly, and as a fixer of defective equipment (the specialty in greatest demand in the hab), she was only about average. The fact that she was not pigeonholed, however, gave her a chance to step back and see the needs of the exploration program as a whole. She thus came up with the Waypoint database as a way of organizing the scientific results that were being generated daily by the specialists into a form that would make them comprehensible and useful to succeeding crews. In short, Heather was one who could see the forest instead of the trees, which is a very valuable skill. As intrepid and good-humored as Cathrine Frandsen, she was a natural choice to lead EVA teams. NASA would do well to recruit her to the astronaut corps.

MDRS CREW TWO

Crew Two was led by Greg DeLory, a physicist with the University of California at Berkeley, and included Don Barker, a flight controller with NASA Johnson Space Center; Gilles Davidowicz, a French geomorphologist; Jon Rask, a biological research engineer with NASA Ames Research Center; Franklin and Marshall University geology professor Andy De Wet (who also participated in the final week of Crew One); independent French-Canadian biologist Fred Janson; and (replacing Professor De Wet during Crew Two's final week) California-based neurofeedback biomedical researcher John Putnam.

Operating from February 21 through March 7, Crew Two expanded the MDRS research program in a number of areas. These were described well by Greg DeLory in his summary report written March 6.

> The science and technology investigations carried out during our time here were as varied as the crew. Gilles brought a unique rover called the "Cliff Reconnaissance Vehicle" (CRV), which scales down the sides of cliffs with the help of a human operator at the top. A camera pointed at the cliff walls during the descent obtains detailed, high-resolution images of the rock strata, aiding in the interpretation of the geologic history of the area. Fantastic demonstrations of this rover concept were performed in "Candor Chasma," a deep canyon a few kilometers southeast of the hab.
>
> One of my pet projects involves techniques to find subsurface water on Mars, using electromagnetic sounding techniques. Materials buried beneath the surface, such as ores, water, and natural gas, can distort ambient electromagnetic fields in predictable ways, detectable from handheld instruments used by scientists in walking surveys. This type of technology may be an important part of future Mars expeditions, as the crew attempts to scout for resources that may be locked beneath the surface—resources that may be key for the mission's very

survival. The team that picked the MDRS site as an analog Mars may have done their job too well—my sounder results showed a very uniform subsurface, and only detected a few possible faults beneath some sandstone.

Fred engaged in research on extremophiles—a unique class of bacteria able to live in extreme environments. Modified versions of extremophiles may be able to someday live and proliferate on Mars, aiding in the successful biotransformation of that planet as part of a larger terraforming effort. Microbes found living in a water-ice mixture in a shadowed region of White Rock Canyon, ~4.5 kilometers southeast of the hab, may fall into this category. Our geologist, Andy, has studied the area from a comprehensive point of view, attempting to relate how the search for past or present life on Mars would fit in with the known geology, emphasizing cooperational efforts between geologists and biologists.

Andy's replacement, John, has measured our brain waves—some of us may have some signs of fatigue, if the larger-than-normal signals in the 1–2 Hz range are any indicator. Clearly, crew mental health is going to be an issue for any long-term Mars exploration mission.

Our most recent investigation here was in the utility of adding sensitive external microphones to the Mars surface suits, in an attempt to enhance the suit user's ability to hear and interpret natural sounds outside, normally muffled by the helmet or obscured by the suit fans. The initial tests of this system were remarkably successful, increasing the communication and safety factors during our last EVA. Such a system should also work on Mars; despite the thin atmosphere, sounds are not below what standard microphones and audio amplifiers can detect.

Crew Two also had the experience of having its commander fall sick for a couple of days, during which time Greg appointed Don Barker to be Acting Commander. Don was a real organizer type, and he took the initiative to put the hab's operational and equipment maintenance and repair procedures into the form of a written station operations manual. Subsequent to Greg's recovery, he made Don Executive Officer (XO), and the appointment of an XO became standard operating procedure for most crews that followed.

ONE GERMAN LEADING FIVE YANKS

Crew Three took over the station on March 10 and operated it through March 24. The commander of the station was Bjoern Grieger, a physicist with the Max-Planck-Institut für Aeronomie, in Germany. He chose for his XO Nell Beedle, a geologist and oceanographer with Fugro Seafloor Research in Seattle. The other crew members included Erik Carlstrom, a geophysicist from Portland, Oregon; Jonathan Dory, a space habitation engineer for SPACEHAB at the Johnson Space Center; and Tiffany Vora, a graduate student in molecular biology at Princeton. Rounding out the crew, serving for one week each in succession, were Sybil Sharvelle, a graduate student in civil engineering at CU Boulder; and Stacy Sklar, a geochemistry graduate student with Northern Arizona University who was also a veteran of Anna Paulson's preliminary MDRS shakedown rotation the preceding December.

Crew Three thus had a German commander leading an otherwise all-American crew, and was split fifty-fifty between men and women. According to the standard wisdom of most NASA human-factors scientists, both of these arrangements are bad ideas for crew composition, and taken together should have been a total mess. But exactly the opposite turned out to be the case. Bjoern and Nell proved to be a perfect match as a leadership team. Bjoern was an easygoing good-natured guy who while very precise in his planning, had no problem with delegating authority. An athletic outdoor woman with a lot of practical skills and much sea time under her belt, Nell was a big-sister type who liked to make sure everything was being done right. (I spent some time with Nell in the Arctic the following summer, and on the basis of our acquaintance, I would say that Nell Beedle is the kind of person who would check the brake fluid before entering a cab.) The crew seemed to take well to being led by this unlikely duo, and bonded very strongly.

This crew had to deal with damage from extreme windstorms, but managed well because of its overall coherence and the particular mechanical skills of Jon Dory, who seemed to be able to fix almost anything.

Another standout crewmember was Tiffany Vora, the biologist, whose lab studies read like poetry. Here is a sample taken from her March 13 biology report:

> I also attempted to stain the supernatant from powdered samples from EVA 27—the Wash Bottom sample and the Hypolith sample. Both stains were unsuccessful due to high background staining. However, I have inoculated both high-salt and low-salt cultures with the Hypolith sample.
>
> Finally, examine the following image. Is it a high-resolution image of the Red Planet, complete with a sunrise and glittering cities?
>
> This picture is actually a poor image taken of the Green Material at 20x. I include it to emphasize an important aspect of nature, an aspect which the crew has been discussing lately. For lack of a better term, a synergy of permutations exists in the universe; atoms, planets, and galaxies all exhibit an overall circular pattern. An eager mind can see the similarities between microscopic particles and our nearest planetary neighbor. I point this out to encourage everyone to notice the beauty and symmetry of nature and of the universe. Take a good look around. Open your mind. Beauty is all around us, and an active, inquiring mind will appreciate the pure elegance of our world.

While many spring 2002 MDRS crews had to deal with large-scale press coverage, this crew had the most, including visits from the *CBS Evening News with Dan Rather,* and the *New York Times*. Veteran *Times* reporter Blaine Harden actually spent twenty-four hours living with the crew, and donned a spacesuit to join them in a full-sim EVA. Harden was positively impressed by both the crew and his experience, and wrote an excellent article that ran on the front page of the *New York Times* on Saturday, March 23.

An amusing sidelight of Crew Three is that its membership was quantized in physical height. Two of the crew members (Jon and Erik) were very tall(~6'6"), three were medium (Bjoern, Nell, and Stacey were all about 5'9"), while the remaining two (Sybil and Tiffany) were quite short, being perhaps about 5'2". This actually had the important positive value of testing the comfort of both the hab and the suits across the range of probable crew morphology. It also led to a delightful photograph in the *New York Times* captioned "Martian Gothic," depicting the giant Jonathan and the diminutive Tiffany, both clad in spacesuits, walking side by side on EVA across the Utah desert.

THE AONIA CREW

Crew Four, running March 25 through April 6, was led by professor Judith Lapierre, a researcher in psychosociology at the University of Quebec at Hull, and included geologist Andrew Hoppin, computer scientist Joel McKinnon, biologist Shannon Rupert, NASA JSC mission operations engineer Jennifer Knowles, and Illinois high school physics teacher Matt Lowry. In addition to carrying forward the ongoing program of field exploration in geology and biology, this crew opened up new fronts for MDRS research in the areas of digital mapping and precise recording of crew activities using webcams and other means to provide abundant data for human-factors researchers. They also tested out a new exploration mobility system: the pressurized rover.

Up through Crew Three, we had been able to rent sufficient numbers of ATVs from local residents to support the mobility of more or less continuous crew exploration activities. With the onset of spring, however, the locals wanted their ATVs back, and this threatened to leave the crew without mobility for most of the time. But the crew turned this problem into a research opportunity by transforming Shannon's Nissan Pathfinder into an analog pressurized rover. With such a vehicle, a crew could travel long distances in shirtsleeves (or at least with spacesuit helmets and gloves off) and then don EVA gear to do outside exploration on foot. This is actually a classic mobility mode for a Mars mission (it's what I proposed for the Mars Direct plan back in 1990, for example), but up until now, we had been avoiding it in favor of our own equestrian ATV system. Crew Four's use of a Pressurized Exploration Vehicle (PEV) gave us a chance to compare the merits of the two approaches. They named their PEV *Aonia,* after a large land feature on Mars, and as this vehicle became such a prominent feature of their rotation, this crew became known as the Aonia Crew.

The Aonia Crew was also the first one to deal with the problems associated with biological life-support systems, in this case a composting toilet called a Biolet.

Restricted to 5.5-kilowatt-hour gasoline generators which burnt out frequently, the MDRS during its first field season was frequently power-strapped. To remedy this situation, we had decided at the beginning of Crew Three to replace the Incinolet (incinerator toilet), which intermittently consumed 3.5 kilowatt-hours, with a composting toilet, which used almost no power. The composter appeared to function well enough through most of Crew Three, although it began to smell a bit toward the end. By the middle of Crew Four, the odor emitted from this now-saturated system had become a serious issue, and the crew was forced to spend increased time and effort to find various ways to induce the goldbricking microbial inhabitants of the Biolet to do their assigned jobs. This unpleasant area of endeavor was to continue with limited success through Crew Five, until Frank Schubert, commanding Crew Six, made a wise decision to remove the failed system altogether.

The most dramatic events of Crew Four, however, concerned human factors.

The commander of Crew Four, Professor Judith Lapierre, is a small person with a big heart and a passionate dedication to the cause of human expansion into space. A French-Canadian, she was a finalist for selection as a member of the Flashline Station summer 2001 crews. She would, in fact, have been selected, had the choice not been blocked by adamant opposition from Pascal, who said she was strongly disfavored by headquarters officialdom at the Canadian Space Agency (CSA). If we chose her, he said, we would offend powerful people. Since I didn't really know Judith then, and there were plenty of other fine people to choose from, I had yielded. But not being one to give up, Judith had volunteered to come to Denver for the summer of 2001, where she worked full-time (at no pay) for Flashline Station Mission Support. In the course of this activity, I and many other Mars Society people came to know Judith, and were all very positively impressed. Accordingly, she was given a commander's slot (the first in our program for any woman, other than in shakedowns) in the MDRS.

The reason Judith was out of favor with certain people at CSA headquarters says a lot about her, and about them. A few years prior she had been a member of an extended space mission simulation conducted in a habitat chamber near Moscow. The crew consisted of a Russian commander and first officer, an Austrian astronaut candidate, and Judith. On New Year's Eve, the Russians had gotten drunk and attacked each other, after which the winner (the commander) had launched an unwelcome physical advance on Judith. A marathon runner, Judith is athletic but very petite; only the combination of the man's drunkenness and her own ferocity in extremis allowed her to drive him off. Afterward, Judith filed a report protesting the incident and demanding action, but high-level Russian and CSA officials dismissed it, as her account threatened to damage international collaboration on the Space Station program. Judith was portrayed as a whiner, and the Austrian, who also filed an objection to the commander's behavior, was mocked by the Russians as a wimp. According to the official story, no incident of the type described by Judith had ever happened.

Except it had. When she came to Denver in July 2001, Judith showed me photographs she had taken of the interior of the Moscow simulation module after the New Year's Eve celebration. The whole place was covered with blood. Judith wasn't lying—the Russian and CSA bureaucracies were lying, and the little spitfire had sacrificed her career as a Canadian astronaut candidate in order to stand up for the truth.

Back to MDRS. At the end of the first week of the Aonia Crew, physics teacher Matt Lowry, who had been an excellent crew member, had to leave. His replacement, a man I shall call Phil (not his real name), was a researcher formerly associated with one of the support contractors at NASA JSC. He was also a major headache. Within a few days of his arrival, he was in severe personality conflicts with every other crew member, including the commander, and would not adhere to crew discipline. He also got on the station com system and started bad-mouthing the rest of the crew over the Internet. This caused great anger among other members of the team.

Judith had no intention of allowing her rotation to be wrecked. She contacted me and asked what she could do. I told her that from my point of view, if someone would not accept crew discipline, they needed to be expelled. But the decision on how to handle things was hers. I would support her in any decision she made.

This was what Judith was waiting to hear. She resolved to expel Phil. The problem, however, was that Phil was physically the most formidable person in the crew, and was manifesting irrational behavior that made it very doubtful he would leave if asked. A further complication was caused by the presence of several reporters for various news organizations, who descended on the MDRS that week. She didn't want to do the expulsion when they were around, and she didn't want to leave Phil alone with any journalists.

What followed was a comic opera in which, for two days running, Judith used various arrangements to send portions of the crew off this way and that, with the main action from a journalist's perspective always with the parts of the crew that did not include Phil.

This strategy almost worked, as Phil was kept separate from three of the four journalists, but finally there was one slipup, and he ended up skunking us to a German newspaper.

Then the press was gone, reducing the issue to managing Phil's physical ejection from the station. Judith judged, correctly in my view, that things would not go well unless there was a clear preponderance of force available, and so she arranged though Mission Support to have one of our local friends, Lamont Ekker, stop by that evening. This proved to be a wise move. Phil did refuse to leave. While Phil was a scientist who liked to work out, Lamont was a husky farmer who didn't need to work out. With his assistance, the other crew members were able to make it clear to Phil that he had worn out his welcome and had no choice but to leave. After a somewhat tense confrontation, Lamont, Joel, and Andrew escorted Phil out of the hab and delivered him to the bus station in Green River.

Obviously, on Mars one could not solve the problem of dealing with an undisciplined crew member by bringing in a native heavy, but encountering the problem in Utah underscores the need to think ahead to keep it from occurring on Mars. Clearly, with better screening we could have avoided this problem in Crew Four, at least for the short time frame of an MDRS rotation. An actual Mars mission will be much longer than two weeks, however, and interpersonal pathologies could develop despite good screening processes. The best way to find such problems in advance might be to put the candidate crew through an extended sim on Earth.

As an aside, I should mention that I have sometimes been asked if I thought that the fact that the Aonia Crew had a female commander had any causal relation to this incident. It may have, but only in the following limited sense. It turned out that Phil, while not having a giveaway name, actually had an ethnic background from a non-Western country where women are held in very low regard. This might have made it difficult for him to function with Judith as a commander. In my view, however, that was his problem. The other men and women in the crew had no problems with Judith.

CLANCEY'S CREW

MDRS Crew Five began on April 7, 2002, and ran through April 21. The commander was Bill Clancey, a NASA Ames Research Center cognitive scientist (actually their Chief Scientist in the area of Human-Centered Computing) who had been part of my Crew Two in Flashline Station on Devon Island during the summer of 2001. Also serving in MDRS Crew Five was another alumnus of Flashline Crew Two, Belgian astronaut candidate Vladimir Pletser of the European Space Agency. In addition to these two veterans, there were four newcomers, including Nancy Wood, a distinguished microbiologist affiliated with Northwestern University Medical School; Andrea Fori, a Lockheed Martin planetary scientist; Jan Osberg, an aerospace engineering Ph.D. candidate with the University of Stuttgart; and David Real, a former *Dallas Morning News* journalist and editor now writing freelance.

Clancey had prepared well for his tour, both personally, with his crew, and with Mission Support, which for purposes of this rotation was implemented by the Northern California Chapter of the Mars Society. Since Clancey is from northern California, and is a member of the chapter himself, this allowed him to meet with his mission support team and obtain a good mutual understanding of what the rules of the game would be. Clancey also had the advantage of prior experience on Flashline, which allowed him to formulate precisely the human-factors questions he wished to investigate.

In terms of its fieldwork, Crew Five was a continuation of the kind of exploration done by the previous crews (assisted in this respect by the donation of three ATVs sponsored by the Kawasaki Corporation, which thus restored our lost equestrian mobility). What was qualitatively different, however, was the degree to which the exploration operations themselves were subject to clinical observation by Clancey. In his reports, he literally clocked through each entire day for two weeks, identifying how all the members of his crew were spending their time. Such a time study can help provide the foundation for determining how crew time could be spent better. Clancey also attempted to do planning, on both a weekly and a daily basis, and then conducted studies comparing the actual activities accomplished with those that were planned.

Clancey is an excellent writer, and his daily reports were filled with many striking observations of both technological issues and crew life. For example in his log of April 15, Clancey writes:

> **1713–1930.** Individual work again. The wind is really picking up. Unfortunately, the hab's weather station has been off-line for over a day. We are unable to restart it and have asked Mission Support for help (there's a phone number, but it's not much use here on Mars).
>
> A few people take the opportunity for showers. Most are working in one place again, distributed almost evenly from the hab computer, workstation area, staterooms, and laboratory. Stopping what I am doing every fifteen minutes to record our activities is not difficult—I find that everyone tends to work in one preferred place or to move between two places. (David moves today between the galley and his stateroom; Andrea moves between the hab computer and her laptop.) This individual stability helps maintain privacy, as well as predictability in sharing the space.
>
> At this time, I send an e-mail to my colleagues at Ames. I want them to begin thinking about designing software that will make the GPS unit fully invisible to an explorer. I don't want to wait for a satellite fix; I don't want to transcribe readings. I don't want to ever know the coordinates at all, let alone have to manually enter or compare them or number waypoints. I want a program to answer questions while I'm on EVA: "Has anyone taken samples near here before?" I also want the program to tell me things like "Warning, you are within 10 minutes of the reserve fuel supply required for safe return to the hab." Being here at MDRS this past week has given me very clear ideas about the navigation assistance and other monitoring required during remote exploration on ATVs. Until now, back at Ames and JSC, we weren't sure what to build; we had the methods but not the requirements. That's why I call what we are doing here "empirical requirements analysis"—finding out what you need to build by doing simulations in the field.

As another example, on April 19, near the end of his rotation, Clancey writes:

> In the afternoon I took a gratifying nap. Looking up through the square portal of my stateroom, I imagined that I was on a spacecraft, landed on Mars. I have been here for months, this is my place, with my bed, clothing, and desk. These are the only people I know. We are here to study the surrounding region of this planet. We must maintain our life support—the power, the water, the greenhouse. We go on EVAs, cook, cleanup, converse, write, read and write e-mail, watch movies. I think again, we six people are alone together for three years. What would that be like? I would prefer to be here with my wife, as three couples. I cannot imagine a monastic existence (a minimal existence) in such close quarters for three years. The Space Shuttle or Station model is not appropriate for such long durations. Of course, sailing expeditions had dozens of people for often over a year. But total isolation for three years was not planned; it was not just a group of six, and not so confined.

The reader will note the comment about taking a nap. As a commander who also was the primary journal writer for the crew's activity (since Crew Five's professional journalist David Real confined himself to doing literary portraits of the individual crew members and other picturesque pieces), Clancey found himself presented with the same problem I had faced in Flashline and MDRS: massive overwork. If you have to both lead the crew and chronicle its activities, you end up putting in nineteen-hour days, waking up at 7 A.M. and sacking out around 2 A.M. After a few weeks, this can be very exhausting. Accordingly, I now believe it desirable that the mission commander and chronicler be two separate people. Unfortunately, this desire is hard to implement, because typically it is the commander who has the most holistic idea of what is going on.

Another innovation that Clancey brought to MDRS was a breadmaking machine. This may sound trivial, but it was actually a very wonderful thing. Fresh bread goes stale after a few weeks, making it impossible to bring ready-made on a Mars mission, and eating crackers all the time gets old. But bread's raw ingredients are indefinitely shelf-storable in compact form. So provided you are prepared to bake it yourself, fresh bread can be made available every day of a two-and-a-half-year Mars mission. It's a major plus for the quality of life.

A final innovation that Clancey brought to our system was a better way of dealing with the professional media. Previous crews had been heavily overrun by TV, radio, and print reporters, with a resulting significant degradation of the quality of the sim. We couldn't simply ban the press, because we needed their reportage to implement the public outreach function of the station. Clancey found a way out of this problem, by insisting that all press visits occur simultaneously on a single "open-house" day toward the end of the rotation. This had the effect of blowing the sim on one day completely but keeping the rest of the time pristine.

I was afraid that the press would not accept this arrangement, but they did. Clancey's open-house day was a great success. In a single day, the crew was visited by crews from ARD TV and RTL TV (German), TechTV (San Francisco), Fox-10 TV (Phoenix), *Der Spiegel* (German magazine), *FACTS* (Swiss magazine), *Dagbladet Daily* (Norwegian), and the *Sunday Telegraph* of London. Most stayed for the entire day, and the resulting press coverage was excellent.

THE EXTREMOPHILES

Crew Six, the final rotation of the spring 2002 season, began April 25 and ran through May 7. The commander was Frank Schubert, and the team included Boulder-based veteran exobiologist and Mars Underground founder Penelope Boston; Steve McDaniel (who had been a biologist on Crew One); and NASA Ames engineer Kelly Snook. Sam Burbank, a multitalented Mars Society member from California who had served in Flashline Station Crew One, and Ephimia Morphew, a human-factors scientist from NASA JSC, rounded out the crew.

This was a very unusual crew, because three of its members (Frank, Sam, and Kelly) were all musicians of professional quality, while the other three were talented amateurs. Accordingly, they decided to form themselves into a rock music group, called The Extremophiles, and test the ability of a crew on Mars to engage in creative collaboration via satellite link with artists on Earth. They had some success at this endeavor. A number of noted musicians, such as Mark Mothersbaugh, became involved on the "Earth" side of the link, and my understanding as of this writing is that the CD containing a selection of the twenty songs they composed is currently being cut and should reach market by late 2003.

This activity interfered some with their field science, but it offered potential for turning more youth onto Mars than might be possible by other means. Anyway, as it was the last rotation of a great initial spring season, I felt we had the right to let our hair down a little. So I gave Frank the green light, and during Crew Six, MDRS rocked.

The Extremophiles' field science program was impacted in a more serious way, however, when early in the rotation, Penny Boston took a nasty spill from an ATV and broke several ribs. Because evacuation would not be possible on Mars, she elected to stay with the crew, conducting her fieldwork through the eyes and arms of Steve McDaniel. This was not an ideal arrangement, since Steve had field and lab work of his own to do, but considering the situation, they accomplished a lot.

In addition to being a professional musician, skilled electrician, and mechanic (the latter being two skills in much demand at MDRS and Flashline), Sam Burbank is also an accomplished independent filmmaker. Accordingly, during his stay at MDRS, Sam filmed a documentary on the station. He edited this himself, and the hour-long show actually aired internationally on National Geographic TV a few months later.

A final interesting sidelight on Crew Six was their diet. Ephimia, Kelly, and Penny were all vegetarians, while Frank, Steve, and Sam had occasionally practiced vegetarian diets in the past. So Crew six decided to go vegetarian. This might have worked, except that some gourmet "Epicurean" vegetarian outfit in Texas offered to sponsor the meals, and the crew accepted. The meals provided turned out to be "macrobiotic," which is to say extreme, even by vegetarian standards. The menu consisted largely of grains, beans, miso soup, wakame sea vegetables, sauerkraut, sushi ginger, and twig tea. In addition to giving the crew gas, this diet also had the effect of requiring (according to Ephimia's precise observations) an average of about 16.5 man-hours per day to prepare meals. This is about ten times the food preparation workload when a normal omnivore diet is employed, and the loss of useful time significantly detracted from crew scientific productivity.

So the lesson here is: no macrobiotic diet on Mars missions.

Frank bore up as best he could under the dietary circumstances. In a series of bold ATV sorties conducted during this rotation, he succeeded in accomplishing an exploratory objective that he had kept in mind since MDRS Rotation One: he found a passable ATV route from the hab through the formidable Skyline Rim. This route would make it possible for future ATV excursions to reach the Cretaceous geologic unit around Factory Butte, which had previously been out of reach.

The impact of the diet was so severe, however, that when the sim ended, Frank ignored the local bar. He got lost in a hamburger stand instead.

I should add that for Frank to have been able to lead this rotation was a miracle in itself. In mid-March, he had been nearly killed when the Cessna 172 he was piloting lost power and then crashed in a downdraft in the middle of Utah's snow-covered Wasatch Mountains. Frank suffered a concussion on impact and was knocked unconscious. He would have frozen to death had not his traveling companion dragged him from the wreckage and kept him awake and thus alive through a night of subzero cold on the mountainside.

His friend's name was Matt Smola.

Frank's leg and wrist were broken, and he was unable to walk for several weeks. But his desire to conduct his musical crew experiment was so intense that he forced his recovery and was ready for action by April 25.

So I guess there really was something to sing about.

THE BIRTH OF EUROMARS

The European chapters of the Mars Society held their first continental convention in Paris at the end of September 2001 and agreed to jointly undertake the project of implementing a European Mars Analogue Research Station, or EuroMARS. A number of candidate locations were identified for the station, including Iceland, northern Norway and Sweden, and the Rio Tinto desert of Spain, and a search committee formed to examine these and other alternatives. A committee of architects was also formed to design the station.

The critical issue, as always, was money. I pledged the support of the American and international Mars Society chapters for the project, and returned from the meeting to the United States to see what I could do.

The strategy that had succeeded for MDRS had been to find an exhibition venue, get sponsors for the exhibition, build a real station to serve as the exhibit, and then deploy it to the field afterward. This plan had worked once—why not try it twice?

During the crisis that had followed the pullout of Disney support for the MDRS exhibit in early 2001, one of the possible alternative exhibition venues I had contacted was the Adler Planetarium in Chicago. Located on Chicago's dynamic waterfront, the Adler was the first planetarium in the Western Hemisphere, and is currently the largest planetarium/space museum in the world. Lucy Fortson, Adler's manager of public outreach, had expressed interest in hosting the MDRS, but the museum could not move fast enough to help us in the summer of 2001. The Adler people had visited the MDRS exhibit at KSC, however, and they were favorably impressed. They told us that if we ever had another such station exhibit, we should keep them in mind.

So I recontacted the Adler and proposed a EuroMARS exhibit for the summer of 2002. They said that if we could provide the exhibit, they would host it, and even offered various forms of assistance during on-site erection that collectively would save us about $20,000. I then got hold of the unions, and the UA came through again. Chicago was their kind of town, exactly the sort of place where future pipefitters might be recruited. They pledged $45,000 to support the exhibit at the Adler.

With this, some smaller pledges, and the offer of free aluminum from Alcoa, we now had a good chunk of the wherewithal to fabricate the Euro-MARS. A chunk, however, is not enough, and none of the donations would come unless the full sum was available. We needed another major donor.

The key break came in January 2002, at the national convention of the Mars Society UK. I was there to speak, and Bo Maxwell, the British Mars Society President, introduced me to multimillionaire entrepreneur Paul Young. Paul had a deep interest in space; he had already formed a launch-vehicle development company called Starchaser Industries, and Bo had recruited him to the Mars Society cause. I had dinner with Paul, looked him in the eye, and told him that if he came through with $90,000, the other donations would fall into line and there would be a EuroMARS. He thought about it for a couple of hours, then came over to Bo's house, where I was staying, and pledged his support. Lacking a readily available bottle of champagne to pop, we had to settle for beer. But under the circumstances, it tasted just as good. The EuroMARS was a go.

The European architects had worked out a design for a station incorporating three decks, instead of the two on MDRS and Flashline. This could be done with only a modest increase in station height, because the headroom above the two decks on the previous stations was more ample than it needed to be. By decreasing headroom from 10 feet to 8 feet, and adding 4 feet to the station height, total available deck space could be increased by 50 percent.

Infracomp had closed up shop in Denver, and we were not entirely satisfied with the work done by BOI, so we decided to build the EuroMARS ourselves. Frank formed a company called Mars Base, rented a lot in Denver, did an engineering design, and contracted out fabrication of a strong steel framework to a local shop. Then, using a combination of hired help and Mars Society volunteers, he erected the framework; added aluminum walls and insulation; built decks, doors, windows; and painted it white. By the end of May, the complete primary structure of the EuroMARS was standing fully assembled in Colorado. A week later, we shipped it off to Chicago, where it was assembled again and had the interior of its lower deck built out to support the summer-long exhibit.

The EuroMARS project advanced a step further when a group of European Mars Society leaders, including Bo Maxwell (U.K.), Charles Frankel (France), and Artemis Westenberg (Netherlands), journeyed to Iceland at the end of June to conduct a search for a station site. Assisted by Iceland Mars Society members Vidar Vikingson and Anna Dis Olafsdottir, and guided by useful advice from the University of Iceland's eminent extremophile biologist Jakob Kristjansson (who subsequently joined the EuroMARS Science Advisory Committee), the team targeted a series of sites in northern Iceland for possible operation. Toward the end of the trip, they were joined by Frank Schubert as well, who went to help assure that the site selected would be suitable for hab construction.

In the end, the site chosen was one in the Myvatn/ Krafla volcanic area of the Icelandic northeast, about 5 kilometers from the village of Reykjahil and 60 kilometers from the Laxamyri/Myvatn airport. The Myvatn site area was devastated by a volcano eruption in 1971, and its vast, nearly uninhabited landscape now resembled a cross between Mars and the Moon. The site featured varied and interesting geology paralleling the volcanic regions of Mars, with numerous niches for endoliths and other forms of extremophile biology. Both the local landowners in Myvatn and the officials in Reykjavik were favorably disposed toward the project, and so at this writing it appears that the EuroMARS has found its future home.

A FINAL BREAK

The other major development of the spring of 2002 was the Mars Society's final break with Pascal Lee.

In mid-2000, NASA management had placed the finances of the HMP under the control of the Search for Extraterrestrial Intelligence (SETI) Institute. (Prior to this, HMP finances had been run, at least in part, through a bank account in the name of Pascal and one other person.) Henceforth NASA funding for the HMP, and such other funding as they knew about, would pass through SETI, who, in return for an overhead percentage on the flow, would use their bureaucracy to make sure that things were being done in accord with NASA's accounting standards.

In November 2001, the Space Frontier Foundation, whose big FINDS fund had been severely impacted out by the dot-com crash, tried to restore their fortunes by holding a special fundraiser at the Playboy Mansion in Los Angeles ("bringing together the worlds of space and soft-core pornography," as one wag put it). In order to motivate people for this worthy cause, the SFF identified four space research projects that would be the beneficiaries of this $1,000-a-plate fundraiser. One of these projects was the HMP.

All that was not our affair. In the HMP advertisement placed in the event program booklet that was distributed to five hundred wealthy individuals, however, Flashline Station was depicted as a SETI Institute-HMP project, thereby implicitly urging donations to the HMP on that basis. This was extremely offensive.

Then relations with Pascal were worsened even more when HMP sent us a bill for $50,000 for services rendered to the Mars Society during the 2001 field season. This bill was wild. For example, HMP charged $2,100 for every Mars Society member who flew to Devon Island from Resolute on a Twin Otter, despite the fact that chartering an entire Twin Otter costs $1,400 and the plane seats seven passengers. HMP also charged $90 a day for any day any Mars Society members stayed on Devon Island (real logistic cost is about $30 a day), even including when they were staying in our station, eating food that we had paid for and using power produced by our generators. In addition, no credit for Mars Society services to the HMP was reflected in the bill, which included the use of the hab, its provisions and generators, and the payment of airfare for several people to the Arctic.

According to Pascal, these charges were necessary to balance the HMP budget. That may be, but by my calculation, actual expenses borne by the

HMP on behalf of the Mars Society were less than $10,000, and if Mars Society expenses on behalf of the HMP were deducted from this, the bill should have been zero.

But Pascal had placed his name on the Mars Society Devon Island operating permits, and was in a position to put a halt to our operations unless he was paid his pound of flesh. So my wife Maggie, who acts as Treasurer for the Mars Society, sent SETI a check for $25,000, deeming it more than ample. At the same time, I moved to have our renewed permits issued in my name, so we could arrange our own logistics and not be subject to further inflated costs charged by the HMP.

Pascal was not satisfied with the $25,000, however, and spread the word in the space community that we were stiffing him on money that he was owed. In March 2002, he called my house and screamed at Maggie, demanding payment. We had received some substantial donations around this time (from the British Lord Camrose and additional funds from Starchaser entrepreneur Paul Young) and so were in position to make payment if necessary. Maggie wanted to pay him off just to get rid of him. At first I was not inclined to do so, but the situation became increasingly acrimonious and I became concerned about possible trouble in the field in the Arctic during the coming summer should Pascal be present nursing this grievance. These concerns were heightened by the fact that Pascal refused my offer to sign a mutual noninterference agreement between the Mars Society and the HMP. The SETI Institute, however, indicated to me that if the bill was paid, they would be willing to put such an understanding in writing (and eventually did so).[33]

So with the money and renewed permit in hand, I decided to pay off Pascal but to tell him in no uncertain terms that the Mars Society would contract for its own logistics services independently in the future.

Pascal denounced this arrangement as "totally unacceptable." But what could he do? I soon found out.

In the beginning of April I received a phone call from Aziz in Resolute, telling me that Pascal was coming to the Arctic at the end of the month "to meet with people." I could see no legitimate reason for him to do this, and plenty of reasons for him not to, as he had recently assumed the demanding responsibility of being the Principal Investigator on a NASA Scout Mission proposal. I therefore suspected that Pascal was going north to try to convince the Nunavut authorities to invalidate our permits. If he could do that, we would have no choice but to use his services again, at rates he had recently set even higher (he was now charging $140 a day for staying on Devon).

I called Chris McKay and requested an immediate three-way telecon to resolve this situation. The telecon was held, and Pascal agreed he would not attempt to invalidate our permits.

I do not know if Pascal honored his pledge after the telecon. However, I found out later from Canadian government officials that, prior to this conversation, Pascal had already contacted the Land Administration (DIAND) in the Nunavut Territorial capital of Iqaluit opposing the extension of our permits, and suggesting that they be consolidated into his. Fortunately, he was unsuccessful. As a result, we no longer had to pay the HMP's grotesque fees and could conduct our affairs without interference from its leader.

We were done with Pascal Lee.

10. A THOUSAND DAYS ON MARS

BACK TO DEVON

As spring 2002 drew to a close, it was time to launch our third field season on Devon Island. In order to save money, we decided to conduct the simulation this year with a single crew that would inhabit the station for three weeks. The crew consisted of me as commander; Nell Beedle, a geologist and oceanographer from Seattle who had served with distinction in MDRS Crew Three; Frank Eckardt, a German-born, British-educated geologist from Botswana; Markus Landgraf, a European Space Agency physicist from Germany; Shannon Hinsa, an environmental microbiologist from Dartmouth University; Emily MacDonald, a Scottish astrophysicist doing graduate work at Oxford; and K. Mark Caviezel, a Pioneer Astronautics engineer.

There were thus seven crew members instead of the usual six. The reason for this innovation was to allow us to operate completely independently of the HMP Camp, without the need for six of the seven crew members to constantly break sim. The seventh crew member, in this case K. Mark, would have the special role of providing all the required out-of-sim support operations, including riding shotgun protection for EVA teams against polar bears. As K. Mark was a former Marine Corps sergeant, and quite adept with firearms, he was highly qualified for this task. He was equally facile with electronics, and proved a very valuable addition to the overall engineering capability of the crew.

Nell served as de facto Executive Officer. I didn't appoint her to the job, she simply assumed it, more or less as I anticipated she would. I did not mind, as her strong organizing personality made her a natural for the job. Nell's forceful streak sometimes came into friction with my own, but her overall merit encompassing scientific, practical, and leadership skills made this a small price to pay for her capabilities. I think she would make a good crew commander.

The rest of the crew were all top-quality as well, but Markus in particular stands out. A meticulous scientist, he was also physically strong, mechanically adept, always good-humored, and willing and able to do practically anything. Whether the task was scratch programming a complex computerized weather station he had never seen before, starting an unstartable ATV, or cleaning out the incinerator toilet, Markus was your man. A real astronaut type. Somewhat younger, Frank was almost as skilled and even better-natured.

The two youngest members of the crew were Shannon and Emily. Both were upper-level graduate students at eminent universities, and each was very focused at conducting investigations in her assigned area. Shannon had transported practically an entire microbiology lab from Dartmouth and organized a group of professors into a scientific backroom to support her work, which was systematic sampling and characterization of the extremophile bacteria in the environment.

Emily had been trained by JPL engineer Mark Helmlinger in the use of a reflectance spectrometer that duplicated the instrument (named MISR) he had flying in polar orbit around the Earth on NASA's Terra satellite. Her mission was to use the instrument to take ground truth reflectance spectra of Devon that could be compared with the measurements taken by the satellite. These would be the highest-latitude measurements ever taken in support of the MISR/Terra program, and, done in simulation conditions, would serve as a powerful illustration of exactly the sort of combined human/robot operations that would be needed to explore Mars. The satellite could see the whole Arctic, but its masters didn't know what it was looking at. Astronauts on the ground have much more limited mobility, but can determine the composition of the local geology with precision. With the help of our measurements, the satellite's controllers would know that this spectrum was dolomite, that limestone, this was breccia, and so forth, and be able to use such data to accurately map mineralogy across much of the high Arctic.

The fact that we were operating independently of HMP imposed some burdens, but it was also a big step forward in realism for the sim over 2001. It created a sense of isolation and self-reliance that will be central to the psychology of an actual Mars mission. Our human-factors data could thus be expected to be improved accordingly.

The season began slowly. As a result of weather, the crew was prevented from flying from Resolute to Devon for several days after our arrival in the Arctic July 6. So we checked into Aziz's South Camp Inn and used the time to train everyone in the use of shotguns and ATVs. We also worked out a written set of mission objectives.

Finally, around 2 P.M. on the afternoon of Tuesday, July 9, word came that the weather on Devon had finally broken. We rushed down to the airport and, in intermittent snowfall, loaded a Twin Otter with one ATV, an ATV trailer, a barrel of diesel oil, three shotguns, some food, and the persons and possessions of K. Mark, Markus, and me. We flew to Devon without incident, and after landing, unloaded the plane with the help of the plane's crew. ATVs and fuel drums weigh about 350 pounds each, and are thus too heavy to lower off the plane with muscle power. So they are dropped to the ground from four feet up. We loaded the trailer with a light load and set off to find a ford to cross the river ("the Lowell Canal") that separates the airstrip from Haynes Ridge, where the Flashline Mars Arctic Research Station (FMARS) is located. Because of all the rain, I was concerned that the canal might be running high, making it necessary for us to cross through several feet of freezing water, as sometimes occurred last year. However my fears proved unfounded, and a ford only a few inches deep was found.

It was an interesting feeling reentering the FMARS, which we had left vacant since the previous summer. I was afraid there might be substantial damage or theft, but the structure was completely intact, and as far as we could see, everything, including a substantial stockpile of food, was still there.

We spent several hours moving our planeload of supplies and equipment from the airstrip to the station, and then had to do some heavy lifting moving the station's diesel generators out of the airlock to their proper positions about 30 meters from the hab. We had brought a charged battery with us from Resolute to start them, as it seemed doubtful whether the generator's own batteries would survive the ultracold temperatures they had experienced during the winter. Before we hooked up the new battery, however, we took bets as to whether the old battery might work. K. Mark said no way. Markus gave it fifty-fifty. I, speaking in jest, said I was certain it would start right up. It did. Everyone was flabbergasted, the spectacle of the generator starting instantly after being exposed to an Arctic winter producing much the same effect as watching the hundred-year-old Volkswagen start immediately in the Woody Allen movie *Sleeper.*

Around 5:30, Nell and Frank arrived with more supplies and two more ATVs. Unfortunately, however, the weather, which was acceptable here, deteriorated in Resolute, shutting down further Twin Otter flights and leaving Shannon and Emily behind for another day. So the five of us worked till midnight stowing supplies and getting the communication system working.

We continued working through the next day, upgrading systems in the FMARS. Then Wednesday evening Shannon and Emily managed to fly in, and the crew was ready to begin operations.

The following extracts from my journal tells the story of the sim.

FLASHLINE STATION JOURNAL, 2002

THURSDAY, JULY 11, 2002

We decided this morning that we were ready to go into sim, so we spent the before-noon hours preparing for our first EVA.

We had left six suits in the station since last summer, and it was questionable how well their backpack electrical systems would survive the winter. It turned out that three suit backpacks were still fully functional, one was partially functional, and two were defunct. So K. Mark and I spent the morning rebuilding the three defective suits, not to their original specification but to the improved design we had since implemented at the Mars Desert Research Station.

Around 1 P.M., we started suiting up the EVA team for their first sortie. The EVA was a four-person pedestrian excursion on Haynes Ridge, consisting of Frank, Markus, Shannon, and Emily, with Frank commanding and K. Mark following out of sim with a gun and a camera to guard against bears and document the trip. Since Nell is a veteran of the Desert Station and I have served before in both the Desert and Arctic stations, this EVA would also have the benefit of providing spacesuit training to almost the entire crew.

The team began to suit up and we had the usual follies of first-timers, resulting in an hour and a half preparation period being required before they made it into the airlock. That's actually quite good: many first-time EVAs on previous crews have taken twice as long. Anyway, at 2:35 P.M. they were out the lock and on their way. They spent two hours patrolling up and down the ridge, walking about 2 kilometers. During this time they observed a variety of different limestone weathering patterns, periglacial features, and fossil and contemporary life forms in association with the limestone. Several samples of rock with lichen were also taken back to the hab for analysis. The crew also traveled a bit down the slope into the crater, where they noted significant amounts of water seepage not unlike the crater water seepage patterns observed on Mars.

Overall, our situation now is excellent. The plumbing in the hab has been fixed by Markus, at least for now. So starting tomorrow we will get the running water system operating. The hab is running fine on one 5.5 kWe generator. Diesel fuel usage is about 12 gallons a day. Crew water use is about 23 gallons a day.

The weather is cold and windy, but we have plenty of surface visibility. So we are planning an aggressive series of survey expeditions, to give the crew, and especially the geologists (Nell and Frank), the big picture necessary to identify potential targets for more detailed study.

FRIDAY, JULY 12, 2002

We had planned today to perform a long-distance motorized scouting EVA, but when we awoke it was snowing and the wind was blowing hard. The snow in the air limited visibility, which would be made worse by snow landing on the spacesuit helmets. In addition, snow cover on the ground obscures all the interesting geology. As a result, we decided to cancel today's EVA and spent the day at the station instead. We managed to get a good deal done.

Nell and I spent the day upgrading the three remaining spacesuits to MDRS standards. Markus replaced the scratched domes on two of the helmets with new ones. He also wrote a report to the Carnegie Institute on progress with the MASSE life-detection experiment.

K. Mark and Frank broke sim to deploy a weather station that has been sponsored to the Mars Society by the Met One Company. It seemed rather complex when we took it out of the box, but Frank showed remarkable aptitude, and ignoring the cryptic instructions that came with the unit, managed to assemble it in less than two hours. Then he and K. Mark deployed it about 200 feet from the hab out on Haynes Ridge. K. Mark did the programming, and as a result, the station is now fully functional and provides precise data on wind direction and speed, barometric pressure, humidity, and temperature. K. Mark is ecstatic about it. He calls it "a meteorologist's wet dream."

Emily assembled the Terra/MISR reflectance spectrometer JPL

sent us. She operated it successfully first inside the hab, and then immediately outside (out of sim), obtaining reflectance spectra of the white standard, snow, rock, and water puddles. Now that we know we can work it, we are making plans to use it outside under simulation conditions when the Terra satellite is overhead. We have a schedule of the overpass times. We will do it first with pure samples, and then with mixed samples reflecting the actual ground types viewed by the satellite. The effect of these measurements will be to provide ground-truth measurements for the orbiter, allowing improved exploration of the entire high Arctic.

Shannon completed assembling her laboratory, which includes a terrific epifluorescent microscope sponsored to her by the Zeiss Company and a molecular laboratory lent by MJ Research. These instruments will be vital to her efforts to identify environmental bacteria. She also tested the water we have been drinking (drawn from the Lowell Canal and stored in plastic jerrycans) for coliform bacteria, and found it to be safe. The water system in the hab is now operational, and she has a test in the incubator to evaluate the water after it has been stored in the hab's storage tanks and run through our pipes. Results of these tests are expected tomorrow.

By mid-afternoon the weather began to improve, and by evening most of the ground snow cover was gone, but this was too late to enable our planned EVA.

Since we have the hab's running-water system operating, we made plans for warm sponge baths or Navy showers for all crew members, as most of us have not bathed since Tuesday. Unfortunately, the warm-water heater appears to be reluctant to do its part. Therefore, as befits the Arctic, the baths will be cold.

SATURDAY, JULY 13, 2002

The weather this morning was overcast, cold, and damp, but it was not raining or snowing. So we finally got a chance to do our first long-distance EVA.

The EVA team was led by me, and included geologists Nell Beedle and Frank Ekhardt, and Shannon Hinsa (a biologist), with K. Mark riding along out of sim to provide armed lookout for polar bears. We started suiting up around 10:10 A.M. and made it out the lock fifty minutes later. We then crossed the Lowell Canal and headed north across the flat plain called the Von Braun Planitia. Our first stop there was a place named Marine Rock, after the U.S. Marines whose paradrops in 2000 brought in the materials that were used to build Flashline Station. Marine Rock is about 3 kilometers from the Station, and we had no difficulty communicating direct line of sight from there to the hab. Frank took GPS coordinates for a waypoint there, and then we pressed on 3 more kilometers north, where we dismounted to climb a large outcrop. From the top of this formation, we had an excellent view in all directions, including a vista of a group of lakes several kilometers farther to the north. We therefore named this hill Lakeview Rock. We also noted that we could obtain satisfactory radio link to the Flashline Station from the hilltop, marking it out as a good potential location for a radio repeater to enable communication across the whole northern sector of our exploration zone. We marked another waypoint on the hilltop and then continued on farther to the north, proceeding to a point identifiable in our landsat satellite images as a location where a small lake empties into a modest river. We took another GPS waypoint here.

So now we had three widely spaced GPS waypoints that correspond to terrain features we can see in the landsat images. This will allow us to impose a map grid on the satellite image points, accurate to within 10 meters. This will be a spectacular aid to surface navigation and exploration.

While obtaining biological samples at the lake, Shannon found an insect exoskeleton. Insects, like most other macroscopic life, are rare on Devon; you generally see about one a year. So finding an insect skeleton was remarkable, and we therefore named the lake Skeleton Lake, and the river draining it the Hinsa River, after Shannon's surname.

We then started our return trip. We first decided to go back by a slightly different route than our outbound leg, but got caught on a kind of peninsula of raised ground surrounded on three sides by steep rocky slopes too dangerous to descend. This reflects a characteristic problem in navigating on Devon that may also exist on Mars. The ground is cut by numerous runoff channels. If you try to stay on the high ground, where you can see where you are going, you run the risk of encountering an impassable channel. If you stay low in the channels, you can't see very far.

So we doubled back and changed our plan to return via our original path. In this decision, we were assisted by a minor innovation that Frank had created the night before in the hab, and that had been duly ridiculed by the rest of the crew as the ultimate in techno-dorkiness. This was the forearm mounting of his fancy path-tracking GPS. With this unique piece of apparel, Frank could read our position in real-time and compare it to our outbound path while riding a moving ATV. Using this device backtracking to Marine Rock was a snap.

SUNDAY, JULY 14, 2002

Last night was Saturday night, and since our work was done by around 9, the crew decided to get together and kick back and watch a movie, courtesy of the DVD player on Emily's computer. The choice of the crew was for comedy. So we watched *Galaxy Quest* and enjoyed it mightily.

The weather today was fairly good, cold but with broken clouds and no precipitation to speak of, so we scheduled another motorized EVA. This one was led by me and included Frank, Nell, and Markus, with K. Mark providing armed escort as usual. The purpose of this excursion was to find good sampling locations for Markus's life-detection experiment. This project is being done in collaboration with the Geophysical Department of the Carnegie Institute, which is developing an instrument called MASSE (don't ask me what that stands for), which will detect either life or fossils by certain biophysical markers. Markus is working with these people to supply samples from Devon Island that will be analyzed by their prototype instrument. To give them the kind of samples they need involves obtaining rock that geologists term "in situ," which means not broken away from the large rock formations that are its origin. They need this kind of rock because they want material that is free from modern biological or weathering action. Such stuff is relatively rare on Devon Island, because of the large amount of glacial activity that has moved much of the surface material around. It was decided that the best bet would be to go to a place where steep canyon walls had laid bare the indigenous rock. The closest such location of my acquaintance is a deep ravine called Devo Rock Canyon.

The canyon is about 2.5 kilometers east of the station, and the route there takes you down the ridge and into the crater, where you pass a pond previous NASA explorers have named Lake Cornell. The drainage from Lake Cornell creates mud traps for the ATVs, which you need to pass by gunning the vehicles and passing through the questionable ground at high speed. We did this successfully, with only one vehicle being briefly stuck. Then we drove up some rising hills to reach the rim of the canyon. The rocks accessible from our south side of the canyon proved unsatisfactory, but looking across the canyon Markus saw something he found much more interesting: Devo Rock itself. Devo Rock is a huge boulder, perhaps 60 feet high, and if it does not strictly comply with the "in situ" criterion, it comes pretty close.

So we backtracked a bit and descended the hills on our ATVs to reach the canyon floor, then waded across the shallow river to reach the rock. While doing this I reflected gratefully on the wisdom of our volunteer spacesuit design team, who had chosen those old-style U.S. Army cold-weather boots as the footgear for our spacesuit simulators. Wearing these we could wade across the ice-cold Arctic stream and not even feel it.

Devo Rock proved fascinating. We found it to be impregnated with 400-million-year-old fossilized algal mats, and around it we found a number of fossil coral-like organisms of similar vintage. Markus climbed about halfway up the rock to get a test sample. First he used a geologist's hammer, but was unsuccessful in breaking off a sample of the type he needed. Then Nell passed him up a 20-pound sledgehammer, and he took a swing. Observing this from the ground, Frank and I didn't like it. He was perched about 30 feet above the canyon floor, and the spacesuits make you clumsy. The big shift in weight involved in swinging the sledgehammer could easily cause a fall, and Devon Island is a real bad place to have an accident. So I called Markus on the suit radio and had him put the sledgehammer down. It was just as well, because his next strategy, using a hammer to knock a chisel into the rock, provided satisfactory results without the sledgehammer's risks.

But Markus still wanted to try the sledgehammer technique, so he came down and attempted it at the base of the rock. He bashed it real hard several times but without results. Then Nell asked for the hammer. She lifted it high and came down hard on the rock outcrop Markus had targeted, once, and then did it again. On her second try the rock broke. Markus bent down and examined the sample and judged it excellent, then got up to leave. "Wait a minute," Nell said. "You mean you're not going to take the sample?" "Oh, no," Markus replied, "the actual sample needs to be taken under much more controlled conditions." Nell was flabbergasted. "I just busted my butt to get that sample, and you're just going to leave it here?" Markus just shrugged, and Nell had to laugh.

MONDAY, JULY 15, 2002

We had a very eventful day today.

The primary activity was a three-person EVA led by me and including Nell and Frank, with K. Mark providing armed escort. We traveled about 6 kilometers southwest from the hab, skirting the outer edge of the crater. In the course of this, we stopped systematically on the outbound leg, recording ten waypoints, including their location, geological characteristics, rock-size distribution, and operational features, such as transversability by ATV, quality of radio contact with the hab, and view of the surrounding terrain. The purpose of this activity is to create a waypoint database that can be referred to by any future crew who might want to know what they can expect to find in different places and make their EVA plans accordingly. In addition, our records of rock-size distribution (in which we estimated the fraction of ground covered at each location by sand, granules, pebbles, cobbles, small boulders, and large boulders) provide a quantitative estimate of the roughness of the ground that can be compared with the coloration on our Landsat satellite images for each point. This comparison will allow us to verify or refute a hypothesis Frank has developed connecting image brightness with surface smoothness, and thus trafficability. According to Frank, other things being equal, the darker pixels on the Landsat images should correlate with rougher ground.

We discovered regions including interesting outcrops, a 400-foot-deep canyon, and an oasis covering at least 10 acres featuring grass, flowers, and even a very stunted tree. There was evidence of an arctic fox den in the vicinity, and we saw tracks of very large animals. There may have been tracks of two different species. One looked as though it might be caribou, except that it was a solitary animal, not a herd. The other looked like bear tracks, except that at 6 inches in diameter they seem a bit small for a polar bear but too big for anything else.

The first five hours of the EVA were thus productive and delightful. On the way back through the crater, however, my ATV was stopped when it broke through a thin layer of dry surface dirt covering subsurface mud. I jumped off and pulled the vehicle out before it stuck hard, but Nell, who was following me, slowed down and her ATV stopped. She did not get off in time, and her vehicle got trapped deep in the mud. She and I tried to push it out but were unsuccessful, and got our feet deeply caught in mud as well. We managed to free ourselves, and then tried to tow the vehicle out with one ATV. This proved insufficient, and in fact the towing vehicle got partly stuck. So K. Mark and Frank went back to the hab to get more rope. Using this, we were able to free the first towing ATV. Then using them to pull together, with Frank pushing and gunning the engine from the side of the vehicle, we hauled out the stuck ATV. The activity was strenuous, and Nell and I both broke sim to remove our helmets and backpacks to allow us to work. In the heat of the fray, Nell's spacesuit dome was smashed and some of the fittings on my backpack helmet connection hose were destroyed.

This was this year's crew's first serious encounter with Devon Island mud, and it was a sobering experience. While the episode was stressful, we worked well together to save the situation. Crew members now have a better idea of how to deal with mud and, I believe, a better sense of themselves as members of a team who can count on one another in a tight spot.

TUESDAY, JULY 16, 2002

Markus and Shannon conducted an EVA to Marine Rock, with me providing armed escort out-of-sim. They successfully collected one of the two sets of pristine samples for the Carnegie Institute MASSE life-detection instrument program. The process was documented meticulously, both by the in-sim EVA team with a digital camera and me with a digital camcorder. I also took a lot of action footage that we can make available to broadcast media who want to do a story on our Arctic Station. The camera is pretty good, but I'm strictly an amateur with video equipment. So if we end up making a movie of our own with the footage we produce here, we should call it "The Blair Mars Project."

On our way back to the Flashline Station, the team did a little scouting and discovered a large interesting outcrop near the Lowell Canal. Markus named the formation Hellas Rock. It's a kick being able to name things, but it is necessary too. Without a name, a geographical feature just sinks into the background. In order to be able to grasp the concept of a thing, talk about it, and hold it in memory, it has to have a name. The first explorers on Mars are going to have a lot of naming to do, but it's a joyous task.

We spent the early evening hoping for sky to remain clear in front of the sun at 9 P.M., because at that time, the Terra satellite would pass overhead. The unlikely coincidence of a clear sky and a satellite overpass is needed to get a good measurement with a JPL sun photometer, but our luck held out and we got our chance. So at 8:40 P.M. Emily suited up and went outside to produce, in sim, the farthest north ground-truth measurement ever taken for the Terra satellite MISR instrument. This data will be sent to Mark Helmlinger at the Jet Propulsion Lab and used to correct for atmospheric opacity distortion of the surface spectra of this region taken by the MISR instrument on the Terra satellite. The result will be an improved orbital survey of huge regions of the high Arctic.

WEDNESDAY, JULY 17, 2002

The primary EVA (EVA 7) was a long-distance motorized excursion consisting of Frank and me, with out-of-sim escort provided by K. Mark. We did a wide circuit, first to the west, then 6 kilometers to the north, then about 2.5 kilometers east, producing fifteen waypoints for the database, with emphasis on rough terrain we believe will correspond to dark spots on the Landsat images. Because we were aiming for irregular ground, we found it, and a lot of this EVA involved tough travel over fields of boulders and broken rocks. Fortunately, since there were only three of us on this excursion, we all were equipped with the larger ATVs, which are much better able to handle this sort of terrain. The worst ground of all occurred near higher elevations, which we had to deal with, since a secondary goal of this excursion was to find hilltops for possible repeater use. Two of those we tested turned out to be excellent spots, located roughly 6 kilometers north of the hab, with commanding views and fine radio reception from the FMARS. But it was a lot of work reaching the summit, especially since in one case the ATVs couldn't make it and we had to hike the last 200 meters to the top.

On one of the hillsides leading to a possible repeater spot, we found several anomalous rocks in close proximity to one another. Situated in the middle of a vast field of broken gray limestone, there was a yellow rock with quartz crystals in it, as well as a limestone rock with an inclusion in it that looked like a stromatolite. There were also some small pieces of a brown mineral that the geologists here have been unable to identify but which Nell, who examined it after we returned samples to the hab, believes to be probably the product of some kind of hydrothermal activity. How these items got to that hillside and why they are all together in the midst of an otherwise undifferentiated limestone field remain mysteries. . . .

We currently have a major problem with the Inmarsat satellite link, which allows us to communicate over the Internet. Starting two days ago, it has been down for up to twelve hours at a time, and it went out again today. We don't know if the problem is with our equipment, the Inmarsat satellite, or ionospheric conditions influenced by solar flares. We have an Iridium satellite telephone and an MSAT phone, but currently we have no estimate on when or if the Inmarsat link will become operational. Until it does, we will use the Iridium as our backup link. This has voice only capability, so we may not be able to send photographs for a while. . . .

FRIDAY, JULY 19, 2002

The weather this morning was cold and windy, but also brilliantly sunny with literally not a cloud in the sky. These clear conditions, combined with the fact that there was a Terra satellite overpass, made today ideal for a reflectance spectrometer EVA.

The excursion was led by Emily Macdonald, and included Nell and Frank, with me serving out of sim as armed escort and photographer. The gear for the reflectance spectrometer includes the instrument itself, a computer control box, a large battery, two tripods, various other items, and some heavy cabling, collectively weighing about 200 pounds, which is too much to carry ourselves. So we loaded it into our ATV trailer, which we hitched to Nell's vehicle, and then set out in convoy.

We first went a small ways into the crater, where we obtained reflectance spectra of various patches of snow. Then we climbed back to the ridge and down the other side to cross the Lowell Canal to proceed north to the Von Braun Planitia.

It was on Von Braun that we spent most of our day, crisscrossing the plain in a series of transects to effectively statistically sample the reflectance spectra from this large, fairly uniform, flat and level feature. Since the Von Braun Planitia is large enough to appear clearly in Terra satellite images, our spectra taken from the ground without any atmospheric distortion can be used to correct the spectra Terra took today from orbit. Now Terra sees most of the Arctic, not just the area within driving range of Flashline Station, but it does not see them as well as we can from the ground. But by working together, we can get Terra's coverage together with our accuracy, and survey the entire Arctic much more effectively than either we or Terra could do alone. This is an example of the sort of combined human-robot operations that explorers will need to conduct on Mars.

The results were good, but I must say that for the EVA team, it was hard work. Even with the help of ATVs and a trailer, lugging all that equipment around and setting it up and packing it away several dozen times really wore them out. But they did it all staying within simulation rules, bolting together tripods, connecting wiring, and operating the control computer, while putting up with the impediments offered by their spacesuits, vision-limiting helmet visors, and thick clunky gloves.

Meanwhile, back at the hab, Markus was writing up his MASSE documentation of yesterday's EVA, and Shannon was working in the biology. The anomalous yellow quartz-containing rock that Frank, K. Mark, and I found on EVA July 17 turned out to have endolithic bacteria inside. Shannon was able to isolate and culture the bacteria in the station's lab, and has imaged them and identified many of their pertinent characteristics. . . .

K. Mark opened up our malfunctioning Inmarsat communication terminal and found tool marks and other evidence that prior users had attempted repair jobs in the field with inadequate tools. Apparently, the unit had shown itself to be defective to prior field users as well.

We have a Twin Otter coming to us from Resolute Bay today with more fuel and a data cable for our MSAT telephone. If we can get that to work, we will be able to send out text data at 2400 bits per second. If not, we should be able to at least fly out some disks loaded with data on the Otter when it leaves.

It's good to have backups.

SATURDAY, JULY 20, 2002

I also had some free time this afternoon, and used it to take a hot sponge bath. This is possible, despite the failure of our water system heater, by warming the water in our kitchen teakettle and then bringing it downstairs by hand. It felt great.

Nell and K. Mark cooked tonight, and instead of our usual fare of canned meat or fish and spaghetti, mobilized 3 pounds of our limited supply of frozen ground beef to cook a huge mess of shepherd's pie. It all went. I have noticed that while there are frequently leftovers after meals are served using canned meat/fish, there are never any leftovers when meals include actual ground beef. This is true regardless of who the cook is, what else is cooked, or the amount cooked. I regard this as a significant observation and commend it to the attention of future Mars mission planners.

Bring beef.

SUNDAY, JULY 21, 2002

The weather last night was overcast, with a descending ceiling, which put my nerves on edge regarding the chances for this morning's Twin Otter resupply flight. But to my great relief, this morning dawned clear, and our Otter carrying fuel, the MSAT data cable, and two Russian TV journalists arrived promptly at 9:15 A.M.

The Otter also carried something else, two pizzas for the crew, care of Aziz Kheraj and South Camp Inn. They were great. Markus in particular appeared transported by the experience, so Nell asked him if it was the best pizza he had ever had. Markus replied no, just the second-best. Nell inquired, What was the best? Markus replied that the best was one that he ate off the stomach of a beautiful woman, many years ago. Well, I guess some things are hard to beat.

So then it was noon and we had our EVA, which today consisted of Markus and me, with the Russian film team following us.

The Russian journalists are Irina and Dmitri, from Russian National Television, or NTV. Their English is imperfect, and I find myself using my fragmentary memory of Russian to fill in the gaps. It is quite remarkable the things you remember after many years of not using a language; scraps of children's poetry, a few lines of Pushkin, and a very odd assortment of words, phrases, and grammatical constructions. Still, they're enough to help.

The Russians brought some great gifts, including chocolate, music, and of course vodka. Irina also presented to me for signature a copy of my book, *The Case for Mars,* translated into Russian under the title *V Zashchitu Marsa* ("In Defense of Mars") by Gennady Gusev, the head of the Russian Chapter of the Mars Society.

In other good news, the MSAT data cable finally arrived. But the bad news is that it is a cable with twenty-five pin connectors at each end, while all of our laptops have nine pin connection terminals. We have a two-ended nine-pin connector from the nonoperational Inmarsat unit, and so Markus and K. Mark took on the job of splicing the two. This was a bit complex, because the two different types of cables use different color wires to mean the same thing, and the same color wires to mean different things. But as I write this, Markus has just shouted in triumph that he has gotten the thing to work. If you get this message, you'll know that he did, and Flashline Station is back on line at 2400 bps, courtesy of our MSAT satellite telephone.

MONDAY, JULY 22, 2002

I think I mentioned that the Russians brought vodka. Well, of all our remaining movies, the best one to appreciate with vodka was *Evolution,* a screwy David Duchovny comedy about alien life forms evolving and propagating in the American Southwest. The film goes very well with hard liquor, and we enjoyed it accordingly.

The EVA today consisted of Frank, Emily, and me, with K. Mark providing armed escort and the Russian film crew following us to get their footage. Our mission was to visit a large number of locations in and around the Von Braun Planitia and its surrounding hills and canyons to obtain rock samples for examination under the reflectance spectrometer back at the hab. The purpose of this is to obtain reflectance spectra for all the characteristic rock types here, thereby providing the data needed by the operators of the Terra satellite MISR instrument to know what they are looking at—not only here, but across much of the Canadian high Arctic as well. We accomplished this goal, acquiring all the samples needed to complete our Terra/MISR reflectance spectrometry ground-truth program. We had to do a bit of hiking to do it though, which once again impressed on me the superiority of human explorers over robots.

We also set up our portable radio repeater on a hill flanking the plain, and used it intermittently as a test for our EVA tomorrow, which aims to travel 16 kilometers from the station and which accordingly will need the repeater to stay in touch.

So now Markus, Emily, and Shannon have the samples they need to complete their research programs, and we still have several days left to us before we need to start pulling out. I therefore intend to make use of the time to do some very long-distance exploration EVAs, going much farther than we have ever gone before under simulation conditions. The plan for tomorrow is therefore a long-range expedition performed in sim by Markus and me, traveling the full 16 kilometers from the hab to the sea. I've never been there before, in or out of sim, and the distance involved is fully twice the 8-kilometer distance record of any previous EVA. Frank has used his Landsat images to map out what looks like a passable route. We'll station the repeater on a hill halfway out to maintain communication with the hab. It's going to be a great adventure.

Tomorrow we push for the sea!

TUESDAY, JULY 23, 2002

The EVA team included Markus and me operating in simulation mode, accompanied by Frank and K. Mark operating out of sim, and Dmitri, from Russian National Television, with his camera. The reason we traveled with two out-of-sim crew members as escorts was the long distance (16 kilometers each way, over rough ground) of the trip and the greatly increased danger of polar bears near the coast. Irina, the producer for the Russian TV crew, wanted to come along too, proposing to ride double on Dmitri's ATV (we only have five), but after consulting with K. Mark and Frank, I ruled that out as too dangerous. Irina was rather upset with the decision, emphasizing that she was a veteran of dangerous reporting in Chechnya, and could handle rough adventures. But I stuck to my guns, and subsequent events made me glad that I did.

This trip started out well enough, with the five of us traveling in good weather over familiar terrain to reach Repeater Hill, 6.5 kilometers north of the station and almost 1000 feet above sea level. We set up the radio repeater on top of the hill, and then went down its back side to follow a streambed that we hoped would lead us to the sea. The merit of following the streambed river valley is that on Devon Island the higher ground is usually covered by the larger boulders that are hard to drive over on ATVs, while the lower ground is filled with the more trafficable pebbles and cobbles that have been eroded off to fall downhill.

Our plan worked for a while, and the streambed took us to a large lake, which unlike most of the little muddy ponds that pretend to the title of "lake" on Devon Island, was filled with a large blue expanse of crystal-clear water. So we named it Crystal Lake. On the far side of Crystal Lake, the river broadened, allowing us to follow it for some distance further. But then the banks started to rise to form small cliffs around first one side then both sides of the river, and as the stream was now too deep to drive in, we were forced to leave the low ground and ride high above its banks along ridges that were strewn with large sharp boulders and cut crosswise by trenches.

The traveling got worse and worse, and on reaching the top of one such ridge about 10 kilometers north of the hab, Frank asked that we call a halt for consultation. The ground was almost impassable, and likely to get worse, he said, because we still had another 150 meters to descend over the remaining 5 kilometers between us and the sea. Looking down at the stream, I saw that it offered no alternative, as it was now narrow, deep, and filled with large boulders, sharp drops, and small rapids. We contacted the hab by repeater, and they informed us that a storm appeared to be moving in, a fact that soon became apparent at our own location as a strong wind picked up and the sky quickly transformed from blue to overcast. K. Mark had driven about 200 meters ahead of the rest of us, and was urging us forward, since the view was great from his vantage point and the ground seemed to be getting somewhat better.

I considered the options. If bad weather had not been in the offing, I would have tried to push on, but when you are on EVA, rain is a threat. The raindrops landing on your helmet faceplate fragment your vision and make it impossible to pick out the safest route to drive over the boulders. So I ordered an abort.

It was good that I did. When K. Mark came driving back to our position, the steering control linkage for his front right ATV wheel snapped. There was nothing we could do to fix this, but fortunately the vehicle remained semidrivable, as the broken wheel just kind of flopped around and, most of the time, liberally allowed the good wheel to determine the craft's direction.

Then, as we started to head back, it began to rain. The view through my helmet deteriorated in no time, and coming down from the boulder-covered ridge, my left wheels went over a depression and my ATV flipped over sideways, throwing me off in the process. So then I was on the ground, with the ATV pitched up a bit more than 90 degrees on its side, threatening to topple over onto me while I tried to hold it back with my right leg. I had been traveling fourth in line at this point, and K. Mark, Frank, and Markus were all ahead of me where they could not see what was going on. Dmitri was behind me, however, and he grabbed the ATV and held it back while I scrambled free. Sometimes having the right kind of press coverage can save your life.

The rest of the drive home was not fun, but at least it was comparatively uneventful. When we got back to the station, we had a debriefing. It is now apparent that the most risky position when we are traveling single file on our ATVs is the last in line, because if you are last, no one is watching you. We have accordingly decided to implement a new procedure whereby whenever we attempt anything difficult or dangerous, we will do it one at a time with all who succeed stopping to look back at the rest until all are across.

Live and learn. Learn and live.

WEDNESDAY, JULY 24, 2002

The weather has deteriorated considerably. It is now cold, overcast, and snowing, with high winds. Consequently, we only did one short EVA today down to the Lowell Canal to give the Russians some film footage. The EVA crew was Irina and I, and we spent most of the time wading through the water, with me answering her questions about the relationship between the search for water on Mars and the search for life.

Given the lousy weather, most of the work today was in the hab. Frank and Nell worked on a summary geology report, which we will transmit tomorrow. Shannon churned out biology reports, and was so successful in this noble endeavor that she overloaded our poor 2400 bps data system with her results. Emily finally pulled together the results of her MISR reflectance spectrometry program, which we also transmitted tonight.

Our plan is to pack tomorrow and pull out to Resolute on Friday. This will give us some margin to catch the planes scheduled to fly south on Saturday. Margin is very much to be desired in this affair, because the next planes out of Resolute won't be until the following Wednesday. We are going to need three Twin Otter flights to pull us all out. This can be quite a trick to manage in a single day.

We've got a lot to do to prepare the hab for winter. The next forty-eight hours will be interesting. If the weather remains bad, they could become tense.

THURSDAY, JULY 25, 2002

We thought last night might be our final one together in the hab, so we celebrated by doing something a bit different for our evening R&R. Instead of our usual 9 P.M. movie, we divided the crew into teams and faced off against each other in a party-type guessing game that Nell introduced to the crew. This turned out to be a whole lot more fun than watching a flick, which supports the suspicion that our modern forms of electronic entertainment may offer less to future explorers than the campfire games of older times.

We had a meeting this morning to plan the pullout. With input from all crew members, Nell had drawn up a list of about thirty tasks that need to be done, and we divided them among us. We also had a meeting to discuss means of improving the station and simulation operations more generally. The compilation of all the suggestions is being written up. They're all good ideas, but some of them may be too costly to implement. Still, because they came from people here who have a fair idea of what is possible, I think we will be able to act on most of them.

We had hoped to get one Otter out today. But crosswinds of up to 30 knots directly perpendicular to the runway made it impossible for planes to reach us.

FRIDAY, JULY 26, 2002

This morning was an emotional roller coaster. When I got up, around 7, the weather was acceptable for incoming Twin Otters, with winds at 6 knots, ceiling at 2000 feet, and visibility at 10 kilometers. I raised First Air in Resolute to give them the good news, and they told me that Resolute conditions were okay for takeoff. No pilots would be available until 8 A.M., however. So assuming that weather conditions remained good, they would take off at 8:30, and we could expect to see our first Otter about an hour later.

This good news got everyone excited; our Twin Otter was coming! But around 8 A.M. the weather deteriorated, with rain and fog limiting visibility, so First Air postponed our flight. Then, around 9, conditions improved, with the rain stopped and ceiling lifted to 2000 feet, with 15-knot winds parallel to the runway (from the west). The clouds broke, leaving us with an unlimited ceiling, with all other conditions fully adequate for Twin Otters. Unfortunately, however, some clown from HMP base camp told First Air the airstrip was socked in with fog and a ceiling at 700 feet, so First Air decided they wouldn't fly "until conditions improve."

I contacted the HMP base camp and had some words with a person there. He said he was looking at a webcam that very instant, and it showed there was fog outside. I suggested that he go outside and look at the sky, which happened to be blue, with plenty of sunshine coming through the broken clouds. He replied that he had two licensed pilots in the tent with him, looking at that webcam, and blah, blah, blah. I suggested again that they go outside and look at the sky.

I don't know whether someone from the HMP base camp corrected their erroneous report, if First Air decided to believe us instead of them, or if the airline got wise through other means, but at 10 A.M. First Air told us they had just launched our first plane, and would launch another within thirty minutes. This set off a mad scramble, as we now had to use our single ATV trailer to rapidly ferry all the gear and science equipment for not just one Otter, but two, down Haynes Ridge, across the Lowell Canal, and up to the airstrip. To add to the fun, one of our ATVs balked at starting for twenty minutes, and then, when the first Twin Otter arrived, it showed up with seven barrels of fuel for our operations next year. So at the same time we were shuttling crew and cargo to the airstrip, we had to also haul fuel drums back, which became progressively harder to manage as the number of available personnel decreased.

But by noon the first and second Twin Otters had come and gone, taking with them Shannon, Emily, Markus, Frank, and the Russians. So now all who were left were Nell, K. Mark, and I, with one ATV and the trailer, and we went through the last tasks of bringing in the generators and power lines, flushing out the pipes, burning the last of the garbage, making the final inspections, and then ferrying our own gear to the airstrip. On the final run done by K. Mark and me back up Haynes Ridge to the hab with two fuel drums, the tailgate of the trailer broke off, sending the drums trundling downhill a ways. Rectifying this situation gave us both some healthy exercise.

Then we brought another load of gear to the airstrip, and I stayed there while K. Mark went back for Nell. It was now about 1:30 P.M. and the weather had gotten bad again, with snow and a very low ceiling. I had the Iridium phone with me, so I called First Air to see if they would be willing to fly out and land under these conditions. They replied that the plane was already in the air, and I should expect to see it within ten minutes. Within a few minutes the plane did appear, but then, to my considerable consternation, immediately disappeared into a cloudbank. I radioed K. Mark back at the hab anyway, telling him to get moving, because (hopefully) our Twin Otter was about to land.

It was really a questionable situation, since the ceiling had now dropped to maybe 150 feet above the runway. But suddenly the Twin Otter appeared in the east, flying lower than treetop level. This is possible to do on Devon Island because there are no trees. Then the plane landed. I felt like cheering.

The 2002 Flashline Station field season was thus brought to a close. Overall, the mission was a complete success. We had operated in full simulation mode for nearly twice the duration of the longest FMARS simulation (Crew Three) of 2001. In addition to conducting a systematic program of field geology and microbiology under simulated mission conditions, the crew had worked successfully with researchers at Jet Propulsion Lab, taking the farthest-north ground-truth measurements ever done for the MISR instrument on the Terra Earth-observing satellite. This was an important demonstration of exactly the kind of combined human/robot exploration operation that will need to be done on Mars. Most important, we demonstrated the ability of the Flashline Station to operate independently of the Haughton Mars Project. Henceforth, the FMARS would be free. Someday, I am convinced, Mars will be as well.

A THOUSAND DAYS ON MARS

Mars Desert Research Station Crew Seven began operations on November 7, 2002, initiating a series of rotations that would run through the following May 11. As of this writing (January 6, 2003), MDRS Crew Eleven is in the station, and together with the FMARS, we have logged 1,207 crew person-days of Mars mission simulation time, with some 95 different individuals having served at least one crew rotation. By the end of the spring field season, we will have had over 140 individual crew members, and total operating time will be approaching the 2,000 person-day mark (person-days, not man-days; roughly a third of our crew members have been female). The current field season includes a fifty-fifty Francophone/Anglophone crew, an all-German crew, a Canadian/Australian crew, an all-NASA crew, and many other diverse teams, with individual crew members drawn from the United States, Canada, Britain, France, Belgium, Holland, Germany, Sweden, Finland, Israel, Ireland, Austria, Australia, and New Zealand.

Many technological experiments are planned. Crew Seven, the Franco/Anglo-American team led by geologist Charles Frankel, demonstrated an improved version of the French Cliff Reconnaissance Vehicle that saw its first tryouts in MDRS Crew Two. Crew Eleven will shortly acquire control of the Michigan Mars rover, an analog pressurized exploration vehicle complete with habitation and laboratory facilities, which will undergo field trains at the MDRS through the month-long Canadian rotation that will end in mid-March. In May, Brent Bos will lead a crew that will test a novel life-detection technology, while during the following field season, Lieutenant Colonel John Blitch will demonstrate a new set of telerobots, and Penny Boston will experiment with an aerial reconnaissance vehicle.

The two largest and most complex experiments currently under way, however, are the Greenhab and the Observatory.

The Greenhab is an experiment in closed ecological life support developed and built by a volunteer Mars Society team, with Gary Fisher of the Philadelphia Mars Society and David Blersch of the University of Maryland being two of the principal figures. Basically, it is a small greenhouse equipped with racks and barrels for hydroponic plant growth, and an automated environmental monitoring and control system powered by an autonomous solar array–storage battery setup. A small flush toilet was installed in the hab, with the idea being that the Greenhab would first be put to work recycling the toilet water (for toilet use only). If it could do that successfully, we would attempt to use it to recycle the hab's gray water (washing and cooking refuse) for washing and cooking use, and if that succeeded, we would use Greenhab treated water to supply our drinking needs as well. This was the first time anyone had ever deployed such a system to a remote field area to provide life support for a space simulation crew. We would be able to find out not only whether the system would work, but its impact on crew time and morale as well.

The Greenhab was installed and activated in mid-October, and ran autonomously with complete success until the arrival of Crew Seven three weeks later. The plants proved unable to cope with the large load of human metabolic wastes, however, and by the middle of Crew Eight, the water in the Greenhab and that recycled to the toilet tank had become brown and smelly to a highly offensive degree. Not only that, the Greenhab duckweed that was supposed to be cleaning the water was actually being killed by it. Accordingly, we directed the crew to reactivate the Incinolet (which could now be easily supported because of the donation of a fine 15-kilowatt liquid-propane generator to our program by the Generac company) for dealing with solid waste, limiting the Greenhab's cleanup job to urine. This has helped somewhat, but the nitrate burden may still be too great.

We are currently contemplating a change in strategy, in which we will direct only the gray water to the Greenhab, disposing of all metabolic wastes by other means. Under these reduced-load conditions, the Greenhab may be successful, and if so it will provide 90 percent of the water-conservation benefit of a complete system (since the gray water flow is much larger than metabolic flow). If all goes well, the Greenhab will then require less burdensome cleaning and maintenance duty, and instead become a pleasant place that provides recreational opportunity for the growing of fresh lettuce and other enjoyable crops. We'll see.

One of the wonderful things about the MDRS environment is its incredible night skies. The station is located in a desert about four hours south of Salt Lake City, reducing light pollution to a minimum, and its dry, elevated (4500 feet) surroundings take advantage of this to produce clear starry skies that are a wonder to behold.

Unfortunately, a crew in sim rarely gets to see any of this, since EVAs are almost always done in daytime, or if at night, under moonlit conditions where their helmet-occluded views reveal only a fraction of the glory that surrounds them.

Accordingly, ever since my tour in MDRS Crew One, it had been my ambition to establish an astronomical observatory at the desert station. The idea was that the observatory would be operated from inside the hab by the crew, who would do their imaging through CCD cameras affixed to the telescope. It would thus be a model for a man-tended extraterrestrial observatory, managed by the crew and responsive to imaging requests forwarded to them by off-site observers through Mission Support. This would allow us to test out the positive and negative morale and operational impact of running an observatory on a Mars mission crew. In addition, it would also provide a means by which the activity of the station could be opened up to participation by such off-site groups as schools and Scout troops, thereby helping to inspire young people to embrace the adventure of science.

In October 2002, I had a speaking engagement in Pennsylvania before the Lehigh Valley Amateur Astronomical Society (LVAAS), one of the oldest and most accomplished amateur astronomy associations in America.

There I had a chance to converse with Peter Detterline, a founding Mars Society member and, as Director of the Boyertown Planetarium, one of the LVAAS's leading figures. I discussed my idea for an MDRS observatory with him. Peter got very excited; he thought it was a great idea, and started working out technical concepts on how it could be done. Within two days he had a list of parts and a budget. Twenty thousand dollars would pay for the whole thing. If I could provide the wherewithal, he would put in the labor to pull it all together.

I took a trip to California, and laid the project in front of our old friend PayPal entrepreneur Elon Musk. Would he fund it? He answered yes. The Musk Mars Desert Observatory was a go. The Celestron Company then pitched in by sponsoring a fine 11-inch computerized Schmidt-Cassegrain telescope, allowing us to splurge on the camera while staying within budget. I scheduled Peter for a crew slot in the late December MDRS rotation 10, and added Debi Lee Wilkinson, another Mars Society astronomer, to the roster as well.

The foundation and dome of the observatory were built during mid-December by Frank Schubert, with substantial assistance by the members of Crew Nine, who broke sim for a couple of days to help out. Crew Ten began operations on December 21. Detterline and Wilkinson would have two weeks to get the observatory up and running.

The initiation of the observatory proved to be a maddeningly complex task, as the two astronomers struggled to make the different electronic systems of the telescope, camera, dome, and hab all work together. Just as the technical systems began to function, they were confounded by freak cloudy skies for several nights running. But in the dark early-morning hours of January 1, they finally managed to take the observatory's first picture, its "First Light" debut.

The image was that of the Great Nebula in Orion, a birthplace for young stars. Exhilarated, the two transmitted the photo with the triumphant cry "First Light! New Stars for a New Year!"

It was a fitting First Light photo for a Mars research station observatory. Stars are the nurseries of life.

11. LESSONS OF THE SIMS

The Flashline and Mars Desert Research Stations have been producing a huge mass of data relevant to the planning of future human Mars missions. Let us now step back and see what lessons can be drawn so far from our experience to date.

The areas of interest break down into three broad categories:

- Exploration tactics and technologies
- Mission technologies
- Human-factors design

These are all interrelated, but it is nevertheless useful for analytical purposes to discuss them separately. In what follows, I will discuss these issues primarily in the context of early human Mars exploration missions, such as that which is being simulated by our stations, rather than that of a mature Mars base.

We start with stating our objective as follows: *The purpose of the mission is to achieve the scientific exploration of Mars.* Accepting that, the implementation of effective exploration tactics is primary. Mission technologies and human factors are secondary, because they must be designed to support the exploration strategy.

EXPLORATION TACTICS AND TECHNOLOGIES

Our strategy is to explore Mars using the best talents of humans aided by the most appropriate tools and robotic assistance. What tactics are required to implement such a plan?

Consider our target. The most obvious fact about Mars is that it is big. Furthermore, it is not only big, it is nonhomogenous, and contains numerous differing objects of interest separated from each other by long distances, and many of these objects are areas of considerable extent in themselves. Therefore, a primary requirement for effective exploration of Mars is mobility.

MOBILITY

There are three realistic mobility modes for human explorers on Mars: They can walk, they can drive, or they can fly. All are potentially important. We consider each in turn.

Walking

When it comes to exploration, there is no substitute for a human walking on Mars. The intimate and flexible direct interaction of a walking scientist with the surrounding environment cannot be equaled or even remotely approached by sequestered humans (in habitats, spacecraft, pressurized rovers, or on Earth) attempting to explore through the use of robotic or telerobotic appendages. This assertion is supported by our experiments with Blitch's telerobots, which show that even with local zero time-delay control, a human attempting to explore a site via telerobotic means is less than one percent as effective as one doing the same job directly. The superiority of the pedestrian scientist over every alternative type of exploratory equipment is so decisive that we state the following as an overall guide to mission design:

All other features of a piloted Mars mission—from the launch vehicles and interplanetary spacecraft, to reconnaissance orbiters, drone aircraft, telerobots, communication systems, navigation systems, power systems, life support systems, in situ propellant, and motorized vehicle systems—serve only one purpose: the delivery of a properly equipped scientist astronaut to a well-selected set of surface sites for direct investigation.

In addition to a unique ability to instantly implement an intelligent and powerful program of sensing and manipulation of the local environment, however, a pedestrian astronaut has another power: he or she can hike. That is to say that the walker's legs can deliver the scientist's brain, eyes, and hands over types of terrain that are impossible for any vehicle to negotiate. Furthermore, on the basis of our simulated EVAs, it is apparent that this ability to scramble over rocks, climb steep hillsides, or otherwise negotiate rough terrain is not something that may only be occasionally useful. In fact, it is frequently essential if scientific objectives such as exploring sites of interest or deploying instruments to necessary positions are to be achieved.

The productivity of the mission will therefore be directly proportional to the ability of crew members on EVA to walk significant distances and durations over rough terrain and otherwise carry out physical activity in the environment. This implies:

1. We need lightweight, flexible, tough, and reliable spacesuits that will allow explorers to hike, scramble, climb, and conduct similar types of outdoor exploration on Mars. The suit must also be maintainable enough to support frequent and repeated use.
2. We need humans to reach Mars in sufficiently good physiological condition to allow them to engage in such an aggressive and extensive program of outdoor physical activity.

These points may seem obvious, but they are both currently being ignored by NASA. There are no significant technology development efforts to create a spacesuit of the type described above. The issue of crew physical fitness is also not being properly addressed. While NASA currently has a huge program in space medicine (the entire Space Station program can be considered research of this type), for the past four decades it is has been myopically focused on the issue of mitigating zero gravity health effects. This is wrong, because as is well established by the failure of both the NASA and Russian programs to find effective countermeasures, long-duration exposure to zero gravity will debilitate the bone and muscle of astronauts to a degree that is unacceptable for an effective Mars mission. For this reason, zero-gravity spacecraft are not the right technology for interplanetary transportation. Instead, interplanetary spacecraft that simulate gravity by making use of rotation to create centrifugal force are required. In order to avoid the appearance of "dejustifying" the International Space Station and its zero-gravity health-effects testing agenda, however, NASA has for the past several decades avoided any artificial gravity research. NASA is spending billions of dollars solving the wrong problem.

Driving

While a good hiker on Mars might be able to cover several kilometers, even the most physically fit astronauts wearing spacesuits as user-friendly as the Mars Society's simulators will not be able transverse on foot the kind of distances required for an effective program of surface exploration. Motorized vehicles will be necessary.

Motorized vehicles break down into two fundamental classes, pressurized and unpressurized, with small and large variants of each type.

Unpressurized vehicles have the advantage of being much lighter than pressurized vehicles, and thus easier to transport to Mars. Their small, light character also reduces propellant consumption, thereby increasing total potential activity. The lightness of unpressurized vehicles also makes it possible for crew members to lift them out of ruts should they get stuck, and allows them to be used to cross rough ground that would be too risky to attempt with heavier systems. An additional advantage of unpressurized vehicles is their ease for launching pedestrian exploration. The driver is already in a spacesuit. If he wants to explore on foot, he just stops the craft and gets off. When he wants to resume vehicular travel, he just gets back on and goes. No pressurization or depressurization cycles are required.

Unpressurized vehicles can either be single-person, like our ATVs, or multiperson, like the Apollo Lunar Rover. On the basis of our field experience, I would argue strongly for single-person systems. The primary advantage of the unpressurized rover is its agility and the informal style of exploration it enables. Both of these characteristics are accentuated in the single-person system, and this is undoubtedly why they have been preferred by manufacturers of all-terrain systems on Earth. Furthermore, the use of multiple single-person vehicles to conduct an EVA in place of a single multiperson vehicle endows the excursion team with a critical degree of redundancy; if one of the vehicles fails, its rider can double up on the back of one of the others. In addition, should one of the excursion vehicles get stuck, the others can be used to pull it free. This latter expedient has been a repeatedly necessary practice in our simulated missions on Devon Island.

Notwithstanding all of the above cited arguments, pressurized rovers have their merits as well. Their primary advantage is their greater duration of operation, and thus greater potential range. An unpressurized rover operation is limited by the endurance of astronauts in their EVA suits. This might be ten hours on the outside, which means a maximum of five hours each way out and back. At an average travel speed over medium-quality ground of 10 kilometers per hour (typical for our ATV EVAs when traveling over well-mapped territory), this would allow unpressurized sorties to attain a maximum range of 50 kilometers from the base. A pressurized rover, on the other hand, contains a living compartment, and this allows the crew to stay out overnight, or perhaps over several nights. Since, to first order, potential travel distance increases in direct proportion to the excursion duration, a pressurized rover with a four-day capacity could go four times as far as a single-day unpressurized ATV excursion, or even farther if we allow the possibility of night travel. Being able to go four times as far means making 16 times more territory accessible. The crew limited to 50-kilometer single-day ATV sorties will have 7800 square kilometers available to explore, but one equipped with a four-day endurance pressurized rover will gain access to 125,000 square kilometers. That's a *huge* increase in mission scientific capability.

So in addition to multiple small single-person unpressurized rovers, a properly designed Mars mission should also include a pressurized rover. Many types of pressurized rovers have been designed, ranging in size from very small-end sport utility vehicles (mini-SUVs) to land yachts. The mini-SUV features a tiny compartment that only allows its crew of two to get out of their spacesuits with difficulty in order to eat cold meals and deal with other metabolic necessities. Other than communication and navigation equipment, the mini-SUV pressurized rover has no scientific gear. It also has no cooking equipment, and its crew sleeps in their chairs. At the other end of the spectrum are the land yachts comparable in size to large recreational vehicles (RVs), equipped with all the comforts of home and a full scientific laboratory. If the mini-SUV is a Gemini capsule on wheels, the land yacht is a Space Station on wheels. The mini-SUV might have an endurance of four days, but the land yacht conceivably could do forty. In between these two extremes are minivan-sized pressurized rovers with intermediate characteristics. Which is the best choice?

Despite its inferior endurance, my vote is for the mini-SUV. The reason is this: Small and spartan as it is, it achieves the fundamental range-extension job required of the pressurized rover. It uses much less propellant than the larger pressurized rovers, and because it is small, it should be possible to liberate it (with a combination of human and ATV power) when it gets stuck. The extended endurance of the larger pressurized vehicle types does not lead proportionally to similarly extended range, because the likelihood of encountering impassible terrain obstacles increases with both vehicle size and proposed travel distance. Thus because of the irregularity of the Martian ground, (and its slow rate of travel), the land yacht is unlikely to succeed as a modality for true long-range mobility. Rather, the land yacht really is a movable base. But if the crew's home base located at their landed habitat is equipped with surface systems providing sufficient reach, there should be plenty of ground for them to explore within the duration of a 600-day sojourn on the Martian surface.

An alternative possibility that might fill the same function as the mini-SUV but weigh even less is the pressurized trailer. This would be a small living compartment that could be towed by a crew member seated on an ATV-style unpressurized rover until the excursion party reached its designated exploration site, after which it would be unhitched. The trailer would then provide a place for the crew to camp out overnight, and then they could continue to explore the selected area with a combination of unpressurized ATV and pedestrian means, or rehitch the trailer and continue their outbound journey. This system would be less comfortable than the mini-SUV pressurized rover, since actual movement would occur in spacesuits, but its light weight and modularity could make it more practical for rugged field use. The Mars Society began considering this idea as a means for augmenting the exploration capabilities of Flashline Station, as it was the only pressurized rover system light enough to fly into Devon Island. For the same reason, it might be the most feasible form of mobile habitat for use on Mars.

The engines that power ATVs and cars on Earth would not work on Mars, because there is no oxygen in the atmosphere to support combustion of their fuel. But one type of automobile propulsion system presently in an advanced state of development for terrestrial use would be very practical for driving Mars cars. This system is the methanol/oxygen fuel cell. Methanol and oxygen can both be readily produced on Mars by processes quite similar to the methane/oxygen combination produced for the Mars Direct mission. When combined in a fuel cell, they produce energy at about 50 percent efficiency (double that of an internal combustion engine), and the water (and CO_2 if desired) exhaust products can be collected for resynthesis back into methanol and oxygen at the base after the conclusion of each rover sortie.

The decision to use methanol/oxygen fuel cell ATV-like vehicles for effective exploration on Mars creates secondary requirements for additional mission technologies. These include methanol/oxygen synthesis technology for the Mars base and spacesuits whose flexibility and other characteristics are compatible with equestrian ATV activity. In addition, because the ATV will be using methanol and liquid oxygen as propellant consumables, it would be desirable for the spacesuits to use methanol/oxygen fuel cells for power and liquid oxygen for breathing gas, and for a means to be provided to transfer these fluids from the ATV to the spacesuit. This would endow the spacesuits with much greater endurance than otherwise possible (including overnight), which in a pinch could allow greatly extended travel range. Such capability could be a lifesaver under emergency conditions by allowing ATVs to be used as lifeboats or rescue craft for pressurized rover missions stranded at great distance from the base.

Finally, the use of ATV-type mobility systems also creates requirements for crew training and selection. It will be essential that crew members be trained in advance to be completely facile with their ATV, to know how to repair it, and to be expert in its use under both nominal and severely off-nominal conditions. A Marsonaut will need to be as familiar with his ATV as a cavalryman of old was with his horse. He'll need to know how to race it or pamper it, heal it, do strange maneuvers and tricks with it, and when necessary, drive it to the limits of its endurance. Like horsemanship, this is not a craft for which everyone has equal talent, so crew members will have to be selected accordingly. Because Mars will be explored in the saddle.

Flight

Mars possesses highly irregular topography, including mountains, chasms, and deeply eroded river valleys. As a result, true long-range (1000-km class or more) mobility can be obtained only through the use of flight vehicles. Even in anomalous conditions where the terrain might be passable for long distances, and assuming a land yacht to be available, the time wasted by such a slow-moving system in reaching the long-range objective would make such an operational mode impractical.

Flight can be accomplished on Mars using balloons, subsonic propeller airplanes or supersonic winged rocketplanes, or rocket hoppers. The prevalence of high Martian winds, however, limits the practical alternatives for human flight to supersonic rocketplanes or ballistic rocket hoppers. Both of these modes require abundant propellant, which is impractical to transport to Mars from Earth. True long-range mobility on Mars thus requires in situ rocket propellant manufacture on a comparable scale to that needed for Earth return transportation in the Mars Direct mission.

The only analog to a rocket hopper capability we have employed in the Mars Society's analog research program is the occasional use of a helicopter. We have not used them much, because they are very expensive, but when they have been used, they have enabled all sorts of movements that are otherwise impossible. For example, MDRS Crew Six used a helicopter to reach the top of Factory Butte, which is unscalable in a spacesuit. A small rocket hopper would offer similar revolutionary capabilities to a Mars exploration crew. Its use would, however, require the availability of considerable amounts of power for propellant manufacture, and depending upon the propellant combination chosen, possibly require access to local water sources as well.

Time

In addition to mobility, another prime attribute required to explore a large planet is time. This makes sense; after all, distance equals speed multiplied by time. But there is more to it than that. Consider first the region available to a Mars mission with a mobility system consisting of ATVs augmented by a pressurized trailer with a four-day field trip endurance. Our range is 200 kilometers, making accessible 125,000 square kilometers of territory. If our mission were of the type using Opposition Class trajectories (twenty-one months in transit out and back on two unequal legs, one month's stay on Mars), we would have an average of *one day* to investigate each of thirty 4200-square-kilometer subregions within our theater of operations. This illustrates the inappropriateness of the Opposition Class trajectory plan for human Mars missions. The alternative to the Opposition trajectory is the Conjunction Class (six months out, six months back, eighteen months on Mars). If this plan is used, we can have about 550 days of exploration, with each day needing to cover about 220 square kilometers. Two hundred and twenty square kilometers is a region equal to a square 15 kilometers on a side. This is roughly the size of the theater of action of the (one-day duration) ATV-equipped crew of Flashline Station, and we have still not explored it adequately after 60 days of simulated crew operations. So even the long duration stay offered by the Conjunction Mission is inadequate to provide sufficient time for a Mars crew to explore their territory. We choose Conjunction, however, because it is much better than the Opposition alternative (Conjunction Class trajectories have less propulsion requirements as well).

This leads to the following three linked statements concerning mission design:

1. Every day on Mars is precious.
2. Every minute of crew time during the surface stay is precious.
3. If you are not planning on spending considerable time on the planet once you reach Mars, you might as well not go.

Statement (2) above is also supported from another point of view by our experience with the mission simulations at Flashline and MDRS. We consistently found that the crews are pressed for time. In order to carry out a program that typically involved one two- or three-person four- to five-hour EVA two days out of every three, and support this level of activity with proper lab studies of collected samples, recording of data, reportage of activity to Mission Support, equipment maintenance, and fulfilling the mundane chores of daily life, our crews of six were generally forced to put in twelve-hour days, with certain individuals (particularly the commanders) frequently needing to work fifteen hours or more.

Crew time is precious. The Mars crew will not suffer from boredom. Quite the contrary, the characteristic problem will be overwork. The Mars crew will not need or desire make-work technologies to give them something to do when they are not out on EVA. They will have plenty to do inside the hab. What they need are laborsaving technologies. This includes devices for dealing with the mundane chores of cooking, cleaning, sewing (yes, sewing), and so forth, as well as those associated with equipment repair, reportage, and lab work.

Given the need to explore impossibly large amounts of territory in a limited time, however, the key area to maximize labor efficiency is that of EVA. In addition to mobility systems, already discussed, the key technologies required to accomplish this objective lie in the area of reconnaissance and navigation.

To use a military analogy, the scientist is the warhead that needs to be delivered to the targeted exploration site; the surface mobility system is the missile that delivers the warhead to the target; the navigation technology is the system that accurately guides the missile to the target; and the reconnaissance system is the means by which the target is selected in the first place based on both its value and its vulnerability (in this case accessibility) to the mobility system at hand.

Reconnaissance

Since even an eighteen-month Conjunction mission stay will be inadequate for the crew to visit more than a few percent of the accessible territory, it is essential that they direct their expeditions to the most important and interesting places. The probable location of such sites of interest can be determined by robotic reconnaissance.

As on Earth, the best means of advance reconnaissance on Mars are flight vehicles, consisting of either orbiting spacecraft or aircraft. Orbiters are optimal for broad area reconnaissance, and can readily be used to map the entire zone of operations with high-resolution photography (10 pixels per meter or better), as well as other forms of remote sensing, including infrared imaging (for determining mineralogy), gamma ray spectroscopy (for surface elemental composition), laser altimetry, and ground-penetrating radar (for detecting subsurface water and stratigraphy). Much of this information can be gathered well in advance of the mission, and will unquestionably be used to help select the landing site.

NASA currently has a good program of orbital reconnaissance of Mars, with two spacecraft, Mars Global Surveyor and Mars Odyssey, active in Mars orbit since 1997 and 2001, respectively. These will be supplemented by the European Mars Express orbiter in late 2003 (carrying, among other instruments, the first rudimentary ground-penetrating radar to Mars), the Japanese Nozomi ionospheric sounding orbiter in early 2004, and the powerful NASA-JPL high-resolution optical imaging Mars Reconnaissance Orbiter in early 2006.

Therefore, of all the technological requirements for a human Mars mission, the need for orbital reconnaissance systems is the one that is currently being well addressed. It should be pointed out, however, that as good as it is, the quality of this orbital reconnaissance will greatly improve globally as soon as human explorers are present anywhere on Mars. This is so because the human explorers on the surface can provide definitive ground-truth information for the imaging, reflectance spectra, and other information acquired from orbit, allowing the satellite's orbital data to be interpreted much more accurately. Such human/satellite synergetic interaction was demonstrated in the field by the crew of Flashline Station and the JPL Terra satellite MISR instrument team during the 2002 field season.

While no drone aircraft have ever been flown on the Red Planet, several serious proposals have been advanced, and one of these, dubbed ARES, is currently under consideration for development and launch as part of the NASA 2007–8 Mars Scout mission. Should the ARES Mars Scout airplane mission win approval, it will provide the first aerial photographic bird's-eye view of Mars. But more important, it will demonstrate a powerful technological tool for use by human Mars explorers. I say more important, because the value of Mars airplane technology is greatly increased if it is used as part of a human Mars mission. The stand-alone ARES robotic aircraft will fly for half an hour, take pictures and other data, and then crash. In contrast, a drone aircraft employed by human Mars explorers can be landed, refueled, and used again and again. Furthermore, the drone employed by human explorers can be piloted remotely in real time, and directed to fly advanced recon along proposed EVA routes and provide high-resolution imaging data of potential surface targets. The information thus obtained can be used to assure the crew that the selected targets are both worthwhile and accessible, thereby minimizing the loss of precious EVA time on fruitless expeditions.

While robotic orbiters and aircraft can serve an important scouting role for human explorers, surface robots cannot. Surface robots are unsuitable platforms for EVA team scouts, because they move much slower than the human explorers and their low-to-the-ground sensing position is worse. Using surface robots as scouts for humans is like using rowboats as scouts for destroyers. That is not to say that surface robots have no value. They can be used in advance of a human Mars mission to gain information about the planet, as the two Mars Exploration Rovers are scheduled to do in 2004. These, however, are not components of an active human Mars mission. If surface robots are to be used in combined operations with humans, their greatest utility is found in their ability to deploy sensors into hard-to-reach places, as demonstrated by the 2001 Crew Three of Flashline Station, when they sent Blitch's telerobots by parachute cord to explore a steep ravine. For this reason, small telerobots are the most appropriate surface robot technology for assisting human explorers.

Navigation

Land navigation is easier than ocean navigation, because there are landmarks. In alien desert landscapes, however, such as Devon Island, southern Utah, or Mars, the landscape, even if varied by irregular topography, can take on a character of sameness. Even if remembered properly, the distinctive features of a topographic object can look quite different when viewed from different angles, and that which was seen from one perspective on the outbound trip can be difficult to recognize on the return. This is particularly true at night (and because of its small moons, nights on Mars will be very dark) or in fog (dusty air on Mars), but it can also be true in broad daylight. The loss of situational awareness of secondary environmental clues concomitant with wearing a spacesuit does not make things any easier. Without artificial aids to navigation, it can be easy to get lost.

At Flashline Station and MDRS our crews use Global Positioning System (GPS) receivers to navigate. So long as the user is well trained and the system does not fail, these can be quite effective, providing navigational accuracy to within a few meters. More expensive units even contain miniature electronic maps, and can be made to depict the progress of the user across the coordinate grid. Such systems would be excellent for use on Mars.

The GPS capability, however, is based on its ability to receive and interpret signals from a large constellation of satellites in orbit around the Earth, which was created at great expense in order to meet the navigational needs of the home planet's 6 billion inhabitants. It could be asking too much to expect that a similar constellation would be installed on the Red Planet for the convenience of the six-person crew of Mars 1. Therefore, somewhat less advanced navigational aids will probably have to be used.

Compasses won't work on Mars (they don't work on Devon Island either), because the planet has practically no global magnetic field. Celestial navigation at night is possible, but in general crews will want to travel by day. The polar orbiting satellites have UHF beacons on them, and with the help of a receiver and a computer, their time of overpass and Doppler shift of their beacon signal can be used to determine both latitude and longitude. The accuracy of this technique is only good to within a few kilometers, however, and it can only be used to determine position during satellite overpass. For short-distance precision navigation in real time, alternative means are called for.

A simple system I believe would do the job would be a set of pole-mounted radio beacons the crew would place in elevated positions at various known locations spanning the zone of operations. These units would be battery-powered, and kept recharged by a small solar panel. They would also contain a radio repeater, and a colored strobe light (with a different color for each beacon) that could be switched on by radio command. Using automated radio direction finding (RDF) gear, an EVA team would be able to precisely determine its location at all times anywhere within the zone of operations by triangulating off the beacons. The RDF-derived location coordinates would be fed into a palm-sized computer, where they would be displayed on a touch-zoom version of the orbital photo-reconnaissance map. The repeaters in the beacons would also enable over-the-horizon communication between the EVA team and the hab. (A communication satellite in Mars synchronous orbit, 17,065 kilometers up, would do so as well but would require much more transmitter power.) The colored lights on the beacons would give instant visual confirmation of a landmark identity, backing up the RDF system, and providing welcome assistance to any EVA team that stayed out overlong and was forced to return in the dark. A set of beacons could also be taken out by EVA teams conducting long-distance expeditions outside of the standard zone of operations, and used to establish a second local RDF triangulation system or simply to mark the return trail home.

While we have not yet employed such beacon systems for navigation, a stationary solar-powered repeater system similar to that described above is used to extend the effective communication range of the MDRS crews. It has its own radio channel, which the EVA teams use only when they wish to speak with the hab. During the Flashline field season of 2002, we employed a portable repeater, which the outbound EVA team would place on hilltops positioned halfway from the hab to their distant objective. The repeater would then be retrieved by the crew during their return trip. This system proved quite flexible and effective.

Scientific Instrumentation

The scientific work of the Apollo missions was limited to sample collection and instrument deployment, with all actual analysis performed by scientists on Earth after the conclusion of the mission. This pattern also holds true for most scientific fieldwork on Earth, with the important differences being that on Earth the scientists themselves typically go into the field and the field excursions are longer. The Earthbound scientists are thus able to refine their investigation plans somewhat as their field season unfolds, and perform some analysis in situ. Nevertheless, the overall pattern is to go into the field for a month or so, gather samples, and then return them to the university lab for extended study and eventual publication. If the lab analysis reveals phenomena of interest in the samples gathered, the field site can be revisited in the following year.

This is not how things will be done on Mars. Because of the tremendous cost of the Mars mission, it will be necessary to maximize effectiveness by equipping the station with a laboratory that will allow substantial analysis of samples while the crew is still in the field. This will enable follow-up of promising leads generated by prior excursions with turnaround times of days instead of years or decades. The data and microscope images taken in the station's lab can also be transmitted to Earth for examination by experts drawn from all the disciplines of the terrestrial scientific community, who will act as the crew's scientific collaborators, through telescience.

What instruments should the station lab carry? The answer to that question depends upon who you ask. The scientific equipment in our stations' labs have included low-power dissecting microscopes, high-power optical and epifluorescent microscopes, a rock saw and polisher, incubators and culture medium, cameras, computers, a glove box, scales, a centrifuge, a refrigerator, heaters, thermometers, and all sorts of glassware and chemicals for various kinds of wet chemistry (ranging from water quality to DNA) analysis. This fairly limited setup has proven sufficient to do a lot of good scientific work, including some microbiology of publishable quality.[34, 35] But some more sophisticated instrumentation would be clearly desirable. I would like a gas chromatograph, an instrument that allows precise determination of the chemical composition of substances. Others might opt for gear that is more elaborate, up to and including ultramodern devices such as a scanning electron microscope.

I will not pass judgment here on particular instrumentation possibilities, other than to note two things. The first is the fairly obvious fact that whatever instruments are chosen will be sharply limited by mass restrictions. It will therefore be a small subset of all the instruments that might be desired, representing a compromise between advocates of various types of scientific investigations. The second point, which I consider less obvious and therefore more important, is that the instruments selected will have to be compatible with the real environment of the hab under conditions on Mars.

This later point is made clear by an anecdote drawn from our experience at the Mars Desert Research Station. The MDRS was equipped during its first field season with a superb epifluorescent microscope lent to the Mars Society by Johnson Space Center. Using this magnificent $20,000 high-powered imaging system, we were able to take excellent photographs of endolithic bacteria and other microorganisms—for about six weeks. After the first six weeks, however, the dusty environment in the MDRS began to take its toll on the sophisticated instrument, and it became increasingly difficult to use. Ultimately, it became nonfunctional. During the second MDRS field season we replaced the epifluorescent scope with an ordinary high-power optical microscope similar to those used by college students everywhere. This sturdy $500 device has performed well without any problems.

On a field trip through the mountains, a mule may frequently prove more useful than a racehorse. Designers of instrument suites for Mars stations would do well to keep this principle in mind.

MISSION TECHNOLOGIES

Beyond those technologies that support exploration directly, there are those that support it indirectly by enabling the mission as a whole. These mission technologies include launch vehicles, interplanetary propulsion, power, communication, and life support. Launch vehicles and interplanetary propulsion are not addressed in our Mars mission simulation program, so I will not discuss them here except to note that our experience indicates that the 8-meter-diameter two-deck tuna can hab design (as conceived for the Mars Direct plan[17]) is very desirable—greatly to be preferred for long-duration livability to the 5-meter-diameter sausage-shaped space station modules. Therefore heavy-lift boosters designed for supporting Mars missions should feature payload fairings of 9 meters or better, so they can meet the requirement to launch such structures.

Experience from our stations does provide insight into the other three primary mission technologies; power, communications, and life support. We address each of these in turn.

Power

Energy is the ability to do work. Power is the ability to release energy per unit of time. Since we want to do a lot of work fast on a human Mars mission, we want to have a lot of power. This pretty much sums the matter up. Nevertheless, we will expand on it a bit to make the matter clear.

Everything in our experience at Flashline and MDRS tells us we want to conduct the Mars mission in a *power-rich* environment. I base this statement on our attempt to do the opposite.

During their first field seasons, we attempted to run both the Flashline and MDRS stations on a minimal power budget of 5 to 7 kilowatts. This was possible, in principle, by adopting strict power economy measures such as shutting off all systems (including the water heater) when they were not actually in use; cooking with only one hot plate at a time; and not turning on any other large power-consuming devices while someone was engaged in cooking.

The problem with this procedure was that it led to power outages whenever it was violated, which, depending on the alertness of individual crew members, could occur frequently. Even when it was followed carefully, it led to all sorts of inconveniences, including greatly extending the time required to cook meals (thus wasting valuable crew time), and forcing a crew member to take a cold sponge bath in order to usefully drain the cold water from the pipes before the newly heated fluid from the just-turned-on water heater reached the showerhead. This tended to discourage the taking of sponge baths with proper frequency, especially in the Arctic. The power shortage also induced us to try unfortunate power-saving expedients, such as the replacement of the effective but power-hungry Incinolet in the MDRS with the stinky low-power Biolet composting toilet. For two-week Flashline or MDRS crew rotations, this is simply an annoying inconvenience. Over the course of an eighteen-month surface stay, however, such a perennial annoyance could have a significant impact on crew morale.

But the situation is more serious than that. While not calculated in the generator power budget for Flashline or MDRS, the energy to drive the ATVs or pressurized rovers needs to come from somewhere. On Mars it would come from propellants manufactured out of local materials using energy from the base power supply to drive the required endothermic chemical synthesis reactions (as, for example, is proposed as part of the Mars Direct plan). Three electric ATVs with 10-kilowatt-hour motors deriving energy at 50 percent efficiency for four hours a day (this is the average daily ATV use at Flashline and MDRS) will use 240 kilowatt-hours of chemical energy per day. If the end-to-end chemical propellant synthesis process is 50 percent efficient, this turns into 480 kilowatt-hours of base generator power per day, or an average of 20 kilowatt-hours around the clock.

If we add this motorized vehicle 20 kWe power requirement to the 10 kWe needed to run one of our stations on a healthy power budget, plus 5 more kWe to cover power needs a real station would have that ours do not, plus another 5 kWe for margin, we arrive at an estimate of 40 kWe as the power needed to properly support a six-person Mars exploration station. This level of power cannot be supplied reliably on the surface of Mars with solar panels. Nuclear power is needed.

Fortunately, NASA has recently indicated that they intend to start a space nuclear power development program. Those who favor human Mars exploration should support this initiative wholeheartedly.

Communication Equipment

Space communication gurus seeking funding like to make a big deal about the advances in communication technology necessary before humans can journey to Mars. But the fact is that current equipment can do the job without difficulty.

Consider the following: Using X-band transmission (NASA's current standard) to communicate over a distance of 2.5 AU (Mars's maximum distance from the Earth), a Mars crew with a 5-meter dish and 200 W of transmitter power (which would require 1 kWe of electricity) would be able to transmit data to a Deep Space Network 70-meter dish at a rate of 2400 kb/s. That's almost ten times as fast as a DSL line, and more than good enough to transmit broadcast television in real time. At Mars's *average* distance from the Earth, this transmission rate would nearly triple. In contrast, our crews have no difficulty flooding Mission Support with reports and digital photographs using a 64 kb/s satellite link. The fact that our stations do not utilize a host of engineering systems that a real Mars craft would have does not undermine this analogy. Engineering telemetry on most spacecraft generally requires less than 100 b/s data rate to support, and even on a more complex piloted system would only need 5 or 10 kb/s at most.

The only way to justify demands for development of new interplanetary communication technology prior to human exploration is to dream up spectacular data-transmission requirements and impose them on the mission. This is sometimes done. I recently read a paper by a com freak who insisted that all EVA team members and vehicles be equipped with high-definition digital video cameras that would be running at all times and transmitting their images back to the hab for instant relay on the Earth. Such an imperative would indeed guarantee telecom researchers a healthy piece of the action from the program R&D budget, but it would serve little other purpose. The crew in the hab doesn't need to see real-time high-definition TV of the EVA, and has better (and more necessary) things to do than to watch it. Furthermore, such local transmission of data floods from the EVA team would impose large unnecessary transmitter mass and power requirements on the mobile group, since outside of the line of sight of the station, all such transmissions would have to be relayed by satellite.

Of course, EVA teams could carry camcorders that automatically record everything they see or drive by, but such tapes or disks would best be returned with the crew for archival purposes. There simply is no need for that much live interplanetary data transmission to support the mission's scientific, engineering, or even public outreach functions. Using just 25 percent of the capacity of the 200 W, 5-meter dish X-band setup described in the previous paragraph, the Mars crew would be able to transmit over 50,000 four-megapixel color digital still photographs every *day* (when Mars is at its *maximum* distance from Earth). There aren't enough planetary scientists in the world to even look at that many pictures, let alone study them. Movies require more data, but using the described setup, the Mars crew could send the full content of Cameron's *Titanic* every day, and still have 85 percent of their capacity free for other purposes.

Life Support

The Mars crew will depend for their lives on an environmental-control and life-support system. The environmental-control system regulates such parameters as cabin temperature, pressure, and humidity. While important, this technology is fairly straightforward, and will therefore not be discussed further here.

The role of the life-support system is to support the crew's metabolism through the provision of oxygen and food, and the removal of carbon dioxide and digestive wastes. There are two approaches for doing this, each with its own school of advocates. One method, known as biological or bioregenerative life support, uses living organisms such as plants or bacteria to generate oxygen and recycle human wastes. Some bioregenerative life-support system advocates go further and recommend using plants and animals to produce food for the crew as well. The other approach, known as physical-chemical life support, uses chemical reactors to split the oxygen out of carbon dioxide and pyrolyze or otherwise destroy digestive wastes. Since there are no plants to speak of aboard, the crew's food supply in such cases must necessarily be transported from Earth.

The bioregenerative approach to life support has a great deal of asthetic charm and is clearly possible. The Earth, after all, is a bioregenerative life-support system and it has sustained us and all previous local life for the past 3.5 billion years without difficulty. Some aspects of the bioregenerative plan—for example, growing food—will clearly be necessary to support human life on Mars in the long run, and Mars colonists certainly will not want to pay the freight for grocery deliveries from Earth. That said, however, our experience in the Mars simulation stations overwhelmingly argues for the physical-chemical approach to life support on early human Mars missions.

The problem with biological life-support systems is that they are unpredictable. The crew is betting their lives on networks of microbes, plants, and animals, which have a very large capacity, and apparent inclination, to behave in ways other than those desired by the life-support system designers. Thus the Biolet composting toilet and the Greenhab waste-recycling systems used on MDRS were both failures—despite previous successful testing elsewhere—because it did not please the microbes involved to work very hard once they were relocated to Utah. The results of the bugs' uncooperative behavior left several MDRS crews living in a very stinky, and possibly unsanitary, situation. In contrast, the simple Incinolet incinerator toilet always worked whenever there was sufficient power to run it.

The unreliability of the biological life-support systems is due to their complexity, which is intrinsic to the system at every level, from the subcellular to the ecological, none of which is fully understood. No one really knows how a cell *works,* let alone a plant or a multiorganism ecology, and there are a million unknown factors that can affect their functions. The only reason the Earth works as a bioregenerative system is that it is huge, and its vast size allows it to buffer itself against disaster. Sets of organisms whose local ecologies collapse are simply replaced by others moving in from the outside, and a new balance is obtained. But in a small system, this does not happen. Even the multi-acre Biosphere II (demonstrated near Tucson, Arizona, in the early 1990s), which was gigantic relative to any greenhouse that might be launched on a human Mars mission, was far too small to stably support all the required biological cycles.

Physical chemical systems, on the other hand, are fundamentally simple, as all they involve is running reactors that drive prespecified oxidation or reduction reactions under controlled conditions. This kind of operation is vastly more predictable than managing living systems. Provided you supply power and regulate the water to have the proper trace chemistry, an electrolysis unit will always produce oxygen from water, and it will do so at a precisely predictable rate. Illuminate a set of green plants with sunlight, and they may produce oxygen, or they may consume it, depending upon how they feel.

Another downside to bioregenerative systems is the amount of upkeep they entail. Maintenance of the MDRS Greenhab in its waste-recycling mode required very substantial fractions (sometimes more than half) of total available labor time, directly impacting exploration operations in a very deleterious manner. What's more, the form of the work (cleaning stinky slime from filters and clogged plumbing lines) was very unpleasant, thereby negatively impacting crew morale as well. A big talking point of bioregenerative systems is that they can be used to grow food, and the activity of raising plants can be a source of aesthetic pleasure for the crew. It is true that when the Greenhab was used for this purpose, some crew members did find it fun, and they enjoyed the taste of the fresh greens so produced. But the overall food yield was tiny compared with the crew's diet as a whole, and the mass of the Greenhab was huge compared with other objects that might have supplied equivalent recreational value—such as a deck of cards, for example, or a board game, a CD containing 500 novels or books of crossword puzzles, a sketch pad, or even a pool table for that matter.

For the Greenhab to have produced sufficient food to materially contribute to the crew's diet, it would have had to be something like 1000 square meters in area (a quarter acre, or a square about 100 feet on a side). While this would still be tiny for a farm—try growing enough food to feed six adults on a plot this size—it would exceed the hab's entire upper deck area by a factor of 50. Such an expansion of habitable space just isn't in the cards for the crew of Mars 1, and even if somehow it could be provided, the work involved in raising sufficient food using techniques of small-scale agriculture would occupy all of the crew's time (or possibly more, as the exhausted crew of Biosphere II found out), negating the fundamental exploratory purpose of the mission. Furthermore, dependence on such an agricultural system would put the crew at risk should a blight destroy their crop.

It's much easier to get your food out of a package than out of the ground. Farming is a noble profession, but it's not the job of a Mars exploration crew. Farmers are typically self-reliant and adept at fixing things, and people of that type are good candidates for Mars exploration missions. But the activity of agriculture won't become important on Mars until much later in the program, when substantial permanent bases are built.

The crew of Mars 1 will get their food the ways city people do everywhere: fresh from the pantry or from the fridge.

HUMAN-FACTORS DESIGN

"The human factor is three quarters of any expedition."

—Roald Amundsen

A central determinant of the success of a human Mars mission will be the morale of the crew. Numerous examples from the history of polar exploration and military conflict show that a crew with the right spirit can pull together as a team and overcome the most formidable odds, while a group lacking adequate morale will frequently fall apart before any real hardship even begins. Napoleon Bonaparte once said, "In war, morale is to material as ten is to one." He knew what he was talking about. In all difficult endeavors that require the cooperative efforts of groups of people, morale is the key to victory.

Part of the effort required to create proper morale on a human Mars mission can be achieved by choosing the right individuals. Clearly, crew members who are tough, resilient, and highly motivated will be less likely to collapse, underperform, or cause problems than those who are soft, apathetic, or given to whining. But under conditions of isolation, even the toughest and most competent group of individuals can fail as a team if their psychological environment becomes so irritating that they start directing rage at one another. I do not wish to overstate this problem; fairly ordinary people such as military conscripts have generally shown themselves able to overcome such tensions to the extent necessary to maintain nominal functionality, with open conflict, mutinies, and so on only occurring in extreme cases. We do not, however, want mere nominal cooperative functionality on a Mars crew. We want a team that sings.

In the ultimate analysis, the ability and responsibility to create a high-spirited "one for all and all for one" crew lies with the crew itself. It lies with the commander, who needs to show the kind of character and leadership that will inspire the loyalty and confidence of the crew members, and it lies with the crew members, who have to show the kind of loyalty and support a commander deserves and needs. It works both ways. Good commanders make good crews. Good crews make good commanders.

That said, it remains the case that mission designers can lower the strain on the crew and thereby substantially contribute to their success by designing psychological conditions that will be as favorable as possible. Observations of crew needs and interactions in Mars analog situations on Earth can do much to expand our knowledge in these areas.

This kind of human-factors investigation has been an important area of research since the beginning of the space program. Perhaps the best compilation of prior observations is presented in the book *Bold Endeavors,* by NASA and U.S. Navy human-factors consultant Jack Stuster.[36] In this seminal work, Stuster examined the history of polar exploration voyages, as well as the workaday experience of the U.S. Navy base at McMurdo Sound, Antarctica, with a goal toward extracting human-factors generalizations that could be used in the planning of long-duration space missions. The populations Stuster examined, however, consisted in large part of groups greatly differing in size, social type, and daily activity from those who would go on Mars missions. It is therefore of great interest to conduct simulated Mars expeditions such as those being done at our stations, to see how well Stuster's generalizations hold up within a mission context that mirrors more closely what a crew will be dealing with on Mars. (In his book, incidentally, this is exactly the direction for further research that Stuster recommends.)

The sections below report some of our observations about various aspects of human factors. Except for the observation concerning boredom, which is diametrically opposed to Stuster's conclusions, and some other material that covers areas that Stuster did not address, most of our findings are in fair to excellent agreement with Stuster's work.

Work

Crews are able to sustain very high workloads. The main potential source of unhappiness associated with work is *not* that there is too much of it but that irrational circumstances (whether accidental or human caused) may be preventing it from being done successfully. In general, the highly motivated crews of our stations are not looking to work banker's hours. They know the importance of the job and they want to get it done. They also want their efforts rewarded with success. Frustration over work can develop if it is perceived that efforts are being directed in an inefficient or unproductive way.

That said, there is too much work to do, and crews will need to pace themselves if they are going to maximize their productivity over the course of their year-and-a-half sojourn on Mars.

Boredom

Boredom is not a problem. Overwork is the problem. There is more than enough to do to keep busy.

Food

Food is very important to a crew, and an extended series of uninteresting meals can erode morale. People tend to look forward to dinner as a major event of the day, both for the food and for the companionship of sitting down together to a relaxed meal and informal conversation. Based on our experience, I would say that feeding the crew with prepackaged meals like MREs or TV dinners would be a very bad idea. Providing the crew with the basic ingredients for meals is a much better plan, because this allows them to use their creativity to concoct a much larger variety of dinners. Rather than being a chore, most crew members found their rotating assignment as cook of the night an opportunity to have some fun while impressing their crewmates. In some cases, a friendly competition developed among crew members in this area. In all cases, the act of sitting down for a group meal was an important bonding activity for the crew, and in instances where there was significant friction among crew members, it served to dispel the tension. I believe, therefore, that having the crew take at least one meal together per day should be mandatory for Mars missions. Because everybody will have lots of work to do that tends to draw them away, this dinnertime break needs to be scheduled.

We have also found that under hab conditions, most people have very broad tolerances for what they will eat (in other words, the fact that you would never order it in a restaurant back home doesn't mean that you won't eat it with pleasure in the Arctic), and will readily accept what other people choose to cook. The only major caveat to this is the distinction between vegetarians and omnivores. For most omnivores, a meal really isn't a meal if there isn't meat in it, which makes meals designed by vegetarians more than a bit of a drag. The vegetarians, on the other hand, won't tolerate meat in their food, which makes cooking pasta or mixed meat and vegetable dishes for a group an annoyingly complicated affair. Basically, I would say that the crew should either be all omnivores or all vegetarians. In no case should macrobiotic vegetarians be included in any crew.

Fresh bread beats crackers or frozen alternatives hands down. Don't forget the breadmaker.

SLEEP

In a small crew such as we will send to Mars, it is important that everyone sack out and wake up on approximately the same schedule. There just aren't enough people to get work done effectively if part of the crew is unavailable during hours when other crew members may need their help. Also, if the crew does not wake and sleep at the same time, they will tend not to take meals at the same time, and this tends to destroy the psychological cohesiveness of the team.

I would therefore *not* recommend a system of watches for the crew. Rather, to the extent possible, everyone should work at the same time, eat at the same time, relax at the same time, and sleep at the same time. Of course, it sometimes happens that this is not possible. In the MDRS, for example, the astronomical observatory required work activity from several crew members at night during hours when others had already knocked off for the evening. This was kept from becoming too disruptive by barring observing after midnight.

What must be avoided at all costs is freecycling, wherein crew members get up and go to sleep at whatever hours they please. We had one instance where a crew commander adopted this practice, and it basically incapacitated the rotation.

A crew must work together as a team, and this is possible only if everyone is sleeping in accord with the same clock.

HYGIENE

In previous centuries, most people bathed infrequently and got by. If we were to guide ourselves by the experience of explorers and mariners of the nineteenth century or previous times, we could dispense with the entire subject of hygiene altogether. We will not, however, be sending nineteenth-century people to Mars, we will be sending twenty-first-century people, and their standards of cleanliness required to support their own morale will be higher.

The question of crew tolerance for reduced levels of hygiene arises from the need to conserve water on a Mars mission. Even with a 90 percent effective water-recycling system, it is imperative to hold water use to an absolute minimum in order to keep mission mass under control. NASA's estimate prior to our work was that a crew water consumption of 8 gallons (32 kilograms) per day per person was an achievable goal. Assuming a crew of six, a 900-day round trip mission, and a 90 percent efficient water-recycling system, this would imply that a Mars mission would need to bring along 17.3 metric tonnes of water to meet its needs. Aside from propellant, this is by far the largest single mass component of the mission, and that is assuming a Spartan water consumption of 8 gallons per person per day, which is a small fraction of the 60 gallons per day the average American uses in ordinary daily life.

So no one on Mars is going to get to take five-minute hot showers every morning. Water conservation dictates that instead they will be limited to intermittent bathing experiences of a much less luxurious type. During Flashline Station Crew Rotations Two and Three of 2001, we imposed a regimen of one sponge bath per crew member once every three days. Along with other water-conservation measures, this allowed us to push average crew water consumption down to little more than 3 gallons per day per person. On the sample Mars mission described above, this would reduce mission water needs by more than 10 tonnes, a savings that would probably cut mission launch costs by 25 percent. I did, however, notice that the limit of one sponge bath every three days made several crew members edgy. Accordingly, in MDRS we changed the regulation for each crew member to one sponge bath every four days plus one Navy shower every four days (that is, a sponge bath alternating with a Navy shower every two days). Employing a Navy shower in place of a sponge bath uses more water, but it allows you to clean your hair in a way that a sponge bath does not. With this moderated routine, crew water consumption was about 4 gallons per day per person, a result I consider satisfactory, as the improved morale justifies the slight increase in water use.

One other comment about washing on Mars missions is in order: It needs to be mandatory. Because of lack of adequate hot water or other reasons at various times in MDRS or Flashline, some crew members elected at certain times not to take their showers or sponge baths as scheduled. This can't be allowed. The problem is not that such crew members begin to smell—sometimes they do, sometimes they don't. The problem is that the crew members in question start to become irritable (this appears to be especially the case with women) without realizing it, to the detriment of the psychological atmosphere of the crew.

Recreation

There is a need for the crew to kick back and relax, both individually and as a group. Individual recreation doesn't need to be scheduled, and requires very little provision from mission planners. By far the most common form of individual recreation observed in our stations is reading. Most crew members seem to prefer to do such reading in public areas, such as the wardroom, where they can snack and talk casually with others while they read. A substantial minority (about one-third) appear to prefer to read privately in their bunks.

Because of the press of work, group recreation needs to be scheduled or it will not occur. At Flashline and MDRS we try to schedule a work stoppage around 9 P.M. Then the crew generally gets together for a movie shown on a computer's DVD player. Which movie to see can be a subject of hot debate, which is generally solved by voting or deal making (We'll see this one tonight and that one tomorrow). The overwhelming favorite type of movie among crewmembers are screwball comedies, such as *Galaxy Quest, Austin Powers, Monty Python and the Holy Grail,* and of course *Mars Attacks.* After a long day's work, people want to get together and laugh. There seems to be little patience for anything serious.

I have observed, however, that crews seem to have the most fun when they forgo technology altogether and engage in such old-time amusements as card tricks, jokes, storytelling, and lighthearted games.

In *Bold Endeavors,* Stuster mentions that the game Risk (a multiplayer board game in which players make alliances and double-cross their allies in an attempt to conquer the world) needed to be banned from an Antarctic research station. We have not had Risk in any of our stations, but I can see why it would be undesirable. In a two-sided game, no one seriously objects that the other side is trying to beat him. But a multiplayer game involving ganging up and betrayal is sure to cause resentments that could become greatly magnified in the confined atmosphere of the hab.

Poker, craps, or other gambling games are a traditional recreation among soldiers and sailors on expeditions, but I have not observed any inclination toward these pastimes among our crews. I think that the reason is that gambling games are an antidote for boredom, and the Mars station crews are not bored. In their recreation time, therefore, the Mars crews are not looking for an intense experience but for activities that dissipate tension.

In getting together at night to watch a comic movie or play a lighthearted party game, or especially to celebrate a special occasion, a bit of liquor is frequently welcome. I say this despite the fact that in my ordinary life back home I never drink hard liquor, and only occasionally drink beer or wine. It needs to be used sparingly, or course, but when handled judiciously a little scotch can do much to improve the atmosphere in the hab. I would therefore recommend that a Mars mission include in its provisions a limited amount of hard liquor.

Most crew members like to have music playing while they work in the hab. What music is played can also be controversial, as tastes vary widely. The solution I adopted was to have individual crew members get to take turns choosing the next CD. This appeared to be a satisfactory compromise. The kind of music that everyone seemed to get a kick out of, however, was live music. During Flashline Crew Three in 2001 we had Christine Jayarajah, a chemist who was also a skilled pianist, in the hab, and whenever she would bang out her Mozart at the keyboard the whole crew would sit up and take notice. There is simply a different quality to music when it is being made in front of you by one of your own.

Mars needs music. I think including someone with musical talent in the crew would be a major plus.

Interaction with Mission Support

Crews tend to bond together against outsiders. This has a positive side, as crew solidarity is a good thing. But it also can be disruptive, since the primary outside group the crew deals with on a daily basis is Mission Support. It is extremely easy for even a soft inquiry from Mission Support to be interpreted by the crew as an unreasonable demand. For example, Mission Support might transmit a message such as "When can we expect today's science report?" The crew has been working hard on the report, but progress has been slowed by the need to deal with intermittent minor malfunctions in various pieces of equipment. The crew looks at the message and thinks: "Those people don't have a clue as to what we are dealing with here!"

Even efforts by Mission Support to assist the crew can be interpreted in hostile terms. An example would be a case where an important subsystem fails, and the crew radios for advice on how to fix it. Mission Support responds by recommending a corrective procedure that the crew has already tried. "Do they think we are stupid? Of course we tried that already!" Or Mission Support might respond with a fix procedure that is impossible to implement for reasons that are not apparent on Earth but quite obvious to the crew. "Get a load of this. Another stupid idea from Mission Support."

One way to deal with this is to have the crew and Mission Support train together before the mission, so they regard themselves as all part of one team. This can help, but the irreducible fact remains that the Mission Support people are safe and comfortable on Earth, while the overworked Mars crew members are risking their lives in a cramped spacecraft 200 million kilometers away. This will not escape the attention of the Mars crew, and there will thus necessarily be a divergence of perceptions between the two groups. Both sides need to be sensitive to this, but the greatest burden for maintaining tact must rest on Mission Support, as they are the party under less stress.

If Mission Support must refrain from giving the crew orders, how is the mission to be controlled? In my view, the only solution for this problem is to place command of the mission with the crew, with Mission Support acting in an advisory capacity. Mission Support will provide satellite reconnaissance intelligence, engineering backup, and scientific collaboration, but the mission will be led from the front. There really is no other way to do it.

Crew Composition

When one is choosing crew members, various attributes come to mind, including character, skills, age, nationality, and gender.

The most important determinant of the value of a crewmember is his or her character. The Mars mission needs crew members who are *resilient,* who do not give in to adversity or frustration, who can maintain their sense of humor not only in the face of danger but of discomfort, and unpleasantness, and stupid screwups of every type. Really, this is key; when things go wrong, you need people who can laugh. Crew members also need to be *rational*—that is able to have their reason rule their emotions, both during moments of crisis and over the long haul. Crew members must be helpful and friendly, and not arrogant, difficult, or egotistical. You don't want anyone aboard who thinks he or she is too important to do menial jobs, or who views himself in competition with other members of the crew for status or renown.

Crew members need to be honest and brave. By brave I do not mean that crew members should be daredevils who take risks for thrills. Such people are to be shunned. Rather, by brave I mean that crew members must be people who do not let fear for their personal safety stop them from doing risky things when accepting risk is necessary, and in particular, do not let it cause them to hold back when an imperiled crew member needs their help. This is essential. Crew members need to trust and respect one another, and it is impossible to trust or respect anyone who is dishonest or cowardly. Crew members must be hardworking and be passionately committed to the mission.

These are a few of the character qualities required of good crew members. One could go on at considerably more length on this subject. To some, such reflections on individual moral character may appear quaint and irrelevant, but they are not. Character is fundamental.

After character, the next most important attribute of a crew member is his or her skill. In *The Case for Mars,* I predicted that the two most important sets of skills for a human Mars mission would be mechanic and field scientist (in that order), and recommended that a crew of four be composed of two of each. Based on our experience in the Flashline and MDRS stations, I would stand by that position. The most important skill in any crew is the ability to fix defective equipment; the crew's best mechanic wins the most-valuable-player award almost every time. The members of the Mars mission will depend for their lives every day on the proper functioning of equipment, and a person who can secure that functioning through repair and maintenance of mechanical, electrical, plumbing, ventilation, communication, EVA, and mobility systems is critical for the safety of the entire team. In contrast, a doctor only comes in handy in the event of a medical emergency, which itself is an event that generally only affects a single individual. After mechanic, the most important skill is that of field scientist (such as geologist, microbiologist, or paleontologist), because the field scientists are the human scientific payload of the mission. Without the field scientists to drive and direct the crew's scientific program, the overall return of the mission would be decreased by orders of magnitude.

So with a crew of four, make it two mechanics and two field scientists; if six people are available, make it three of each. No one should be assigned a slot with doctor as *primary* assignment, since medical skill is only valuable in the event of a contingency. Nor should any person be selected whose primary job is to perform medical experiments on other members of the crew. Such people are very irritating and would probably end up out the airlock. To meet the needs of medical contingencies, however, it would be good if one of the field scientists or mechanics was a person who had practiced medicine in the past and would be prepared to do so if such necessity arose.

There is a need for a mission commander, but because of the shortage of hands available, that person will also have to be one of the mechanics or field scientists.

Mechanics and scientists: In *Star Trek* terms, Scotties and Spocks. These are the skills needed to explore Mars.

As for the rest—nationality, age, and gender—I believe a great deal of latitude is possible. Classically, the favored crew composition would be a majority-male group from a single nation, with one of the oldest males in command. It will be observed that this plan is also the description of a Neolithic hunting party or a Bronze Age warrior band, which is not to deride it but to establish its heritage. Because this has been the traditional social organization of human expeditionary groups for the past 40,000 years, people are most comfortable with it, and one should exercise due caution in departing from it.

For example, consider the issue of age. Youngsters are typically not found in command of expeditions, because experience and wisdom are desired attributes of a commander, and age is generally a requirement for their acquisition. But let us assume that in a particular instance a man of twenty-two could be found who was, in fact, equal in these qualities to an alternative commander candidate who was fifty. It would still be very questionable whether it would be wise to assign the commander's position to the young man. The commander's job is the hardest job in the crew, and it would be much tougher for the young man to do it, since he would have to battle perceptions of immaturity and the human instinct to assign the role of commander to a father figure. It would therefore *not* be a favor to the talented young man to place the burden of command upon him.

The same logic holds, but with lesser force, to the possibility of a woman commander. Given equal talent and experience, the woman would have to work significantly harder than a male alternative to fill the role of commander. This is not to say that there cannot or should not be a female commander on a human Mars mission. But such a choice would need to be justified on the basis of clearly superior ability. (Several women of my acquaintance say that this will not represent a problem.)

With respect to the crew as a whole, I favor a spectrum of ages, with the average age perhaps around forty. Having a younger person or two in the crew brightens the atmosphere, but you really don't want a crew dominated by those under thirty-five, as they can be quite competitive. In contrast, people over forty-five generally have already established their careers and identities—they are not out to prove themselves, and so are better at creating a cooperative atmosphere.

All-male crews are possible, but given the importance of establishing the Mars program as a fundamental human endeavor, and not a mere boys' game, I believe it is imperative that the expedition to Mars be comprised of a mixed group of men and women. Based on our experience in the analog Mars stations, I would say that if you have one woman in a crew, you want to have more than one. Otherwise, the single woman in an otherwise all-male crew is left in an uncomfortable position. Therefore, assuming six crew members (as we have at our stations), I believe the optimal crew composition is either four men and two women, or three of each.

A Mars exploration craft is a bad place for sexual competition. Therefore, a mixed-gender crew should probably be composed of either couples or people who are firmly attached to life partners who are not on the mission. The individuals chosen for such a crew must be men and women who can regard members of the opposite sex as people and not objects. A Mars crew cannot be a singles bar. It needs to be a band of brothers and sisters, true comrades embarked together on a great adventure.

The history of Arctic exploration has shown that nationally homogeneous crews have the edge over multinational teams, because they tend to be more cohesive when the situation becomes difficult. A very undesirable crew is one split evenly, or nearly evenly, between two nationalities that employ different languages, as clique formation can readily develop. We have not observed this pathology in our stations, despite having several crews that were vulnerable to it (Flashline Crew Seven in summer 2002 had two Germans and four Anglophones, MDRS Crew Seven in fall 2002 was divided three-three between Francophones and Anglophones). These crews only ran two to three weeks, however, and neither faced any sustained crises. In general, I think it is safe to say that the viability of an international crew depends heavily upon the particular individuals chosen. If your crew members are people with strong flag-patriotic identities, then a single nationality crew is greatly to be preferred. If, on the other hand, the crew is composed of people who locate their identities as cosmopolitan world citizens, a multinational crew may work out just fine.

The above statements about nationalities, age, and gender of crew members are, however, all generalities that are easily undermined by variation in individual personalities. Moreover, it is also the case that some people might operate very well in combination with certain crew members but badly with others. Notwithstanding all the generalizations of human-factors research, the best guide to future performance of a crew is its past performance. The only way really to be sure about a candidate crew is to test it in the field. In my view, the way to choose the crew of Mars 1 will be to select five good candidate crews, and test them simultaneously in one-year field exploration missions in Mars analog stations. The crew that works out best as a team wins.

The Strongest Link

Professional pessimists and others wishing to throw cold water on the prospects for human Mars exploration have sometimes made the claim that failure is almost inevitable, because despite all the hubristic technology that space engineers might bring to bear, "the human psyche will be the weakest link in the chain."

The experience of our crews does not support this. On the contrary, time and again, crews have been presented with system failures that threatened to end their mission. In all cases, they have managed to find ingenious technical fixes that saved the day.

It is true that problems are sometimes caused by the crew members themselves. But the combined good within them has always outweighed the bad and allowed them to prevail. Perhaps the most emblematic human-factors lesson to be drawn from our program can be extracted from the story of its birth, the crisis-wracked construction of Flashline Station.

Everything did not go right on Devon Island during the summer of 2000. Neither, however, can we expect everything to go right on the first human mission to Mars. The military has a saying: "All plans fail upon contact with the enemy." In the wild Arctic, all plans fail on contact with reality. The same will be even more true on Mars. When venturing into the unknown, the unexpected *will* happen. But a resourceful crew can deal with it.

On the piloted Mars mission, the human crew will be the strongest link in the chain.

12. WHERE DO WE GO FROM HERE?

Come, my friends,
'Tis not too late to seek a newer world.

—Alfred, Lord Tennyson, *Ulysses*

Mars is within reach. Despite the recent *Columbia* accident, and despite all the technical problems that could be named, or imagined, the fact remains that we are much better prepared today to send humans to Mars than Americans were in 1961 to send men to the Moon—and we were there eight years later. It is the decision to act that is key.

In May 1961, when Kennedy made his speech committing the nation, the United States had fifteen minutes of suborbital manned spaceflight under its belt. We did not even know if humans could eat in space. Today, the United States, as well as Russia, has flown many missions in which individuals have spent six months (the time required for Earth-to-Mars cruise with present-day propulsion) on-orbit in zero-gravity conditions that are physiologically much harsher than those that would be experienced on a properly designed artificial-gravity Mars spaceship.

We are also much better prepared scientifically. There are a number of areas in which it would be useful for robotic probes to provide further advance information before humans walk on Mars. The United States has already flown sixteen probes to Mars, however, eleven of which have been successful, and as of this writing, two more are being readied for flight. These have produced a wealth of data that can be used to prepare a human exploration program. In contrast, in 1961 it was not even known if the Moon had a solid surface, and some were predicting that the Moon landers would simply sink into the dust and disappear upon touchdown.

We are better prepared militarily. Despite the tense international situation of the early 2000s, the United States faced much greater threats in 1961. In 1961, a hundred Warsaw Pact divisions stood on the Elbe River waiting to advance, backed up by 10,000 Soviet nuclear warheads. The current batch of international terrorists and their allied states do not compare with this. In 1961, it was possible for some people to complain about spending precious funds on Moon shots when the resources in question could be much more usefully spent building hydrogen bombs. One does not hear this argument today.

We are better prepared financially. Humans-to-Mars is a problem of lower technical difficulty for our society today than manned flight to the Moon was in 1961, yet we have three times the GNP, in real inflation-adjusted dollars, than John F. Kennedy's America. Were we to do this with an international partner, such as Russia, Europe, or Japan, each of them for differing reasons unavailable as potential teammates for Apollo, ten times the material base could be available to support the Mars program.

Finally, we are better prepared socially. We still have social problems in contemporary America, and the West more broadly, and we always will. Yet such social problems as we face today do not compare with those we had in the 1960s, when people were burning down cities in this country in frustration over poverty, racism, and lack of opportunity.

In short, from every point of view—technical, scientific, military, financial, and social—we are much better prepared today to launch a humans-to-Mars program than the America of 1961 was to begin Apollo, or any society in human history has ever been to launch an age of exploration at any time in the past.

So if not now, then when? If not us, then who?

Mars today offers us a historic opportunity to open an unlimited future. By establishing the first human foothold on Mars, we can take the decisive step toward transforming humanity into a spacefaring species. But opportunities need to be seized.

It is possible that a humans-to-Mars program could be initiated by the political class. The West could start such a program to show the superiority of a civilization based on reason to one that denies it, thereby helping to further the struggle against Islamic terrorism. Or the United States could send humans to Mars to show up the Chinese, or the French, or simply to provide some much-needed flash to an otherwise drab presidency.

But what then? If the Mars program is merely a tool of the political, it will be dropped by them, just as Apollo was dropped, when its services are no longer required. This is unacceptable. The true historical purpose of the Mars program is not to meet the contingent needs of the present but to unleash the potential of the future.

We need to go to Mars not to show off, or enlighten, or even to merely explore. We need to go to Mars to make the Red Planet a new home for life and humanity.

This is an impractical goal in the eyes of those who only consider the needs of the present, but a necessary one. If it is not achieved, then the human race will remain a single-planet species, and new nations that are waiting to be born on Mars and many worlds beyond will remain unborn.

Such a grand objective cannot be achieved by those who occasionally support it for other purposes. It can only be realized by those dedicated to the idea itself. With the help of significant education and exhortation, the politicians may yet get a crew to Mars. But the Red Planet will not be pioneered unless there is a force in being committed to the idea and prepared to make it stick. That is why we need the Mars Society.

There are hundreds of millions of people who share the belief that it is essential for a positive future that humanity expand into space. If we band together, we can make it happen.

THE NEXT STEPS

Measured on the scale of grand human accomplishments, the Mars Society's achievements to date have been modest. Yet all great journeys must begin somewhere, and we have made a start. Our Mars analog research station program has been successful. We are now gathering financial and technical resources for our second project, in which we will, for the first time, test out the effects of Mars gravity on terrestrial mammals. Currently planned for 2005, the Translife Mars Gravity Mission will carry a crew of mice for fifty days in a small rotating spacecraft in Earth orbit. During this time, the mice will be allowed to reproduce and their young to grow up, after which they will be recovered for study. The data gained from this experiment will be key to understanding the effects of Martian gravity on human explorers and future colonists.

NASA has never flown an artificial-gravity mission of any kind, let alone one focused on the effects of Mars gravity on higher lifeforms from Earth. The Translife Mission will be a first in another respect: it will be the first time that mammals have ever reproduced in space. In financial terms, the mission will also be a major step for the Mars Society. The Mars station program, in round numbers, has been a $1 million project. Translife, our first spaceflight mission, will cost around $10 million. Should we succeed, however, we believe that we will be then poised to launch space missions in the $100 million class, enabling the first privately sponsored robotic missions to Mars. These, in turn, could earn us the credibility required to mobilize the billions of dollars needed to start privately funded human Mars exploration, and ultimately settlement.

On the evening of July 28, 2000, as I gathered my thoughts for the speech commissioning Flashline Station, I had a vision of what it meant, and where it might lead. I penned it down, and a few hours later managed to suppress my cough well enough to deliver it to the assembled ragtag group of volunteers and Inuit who had beaten the odds to build the station.

I think it is fitting that this work be closed with what was said then.

SPEECH AT THE OPENING OF FLASHLINE STATION

We are here to commemorate the opening of the Flashline Mars Arctic Research Station, a place where we will first learn how to train for Mars missions, and then, I believe, ultimately be used to train the first crews for Mars.

This, then, in a real sense, is the place from where the mission will be launched.

But it is also the launching point in more ways than just that.

Because what we have done here this summer is launch a movement. We have shown the world that the Mars Society, the people who believe humans should reach out to Mars, is ready to back words with deeds. To pull this thing off after the disaster of the failed paradrop, in which we lost our fine fiberglass floors, our precisely engineered trailer, and our crane, required imagination, courage, skill, hard work, teamwork, and above all, grit and determination.

These are the qualities that will get us to Mars.

So when you see the ugly but serviceable wooden floors that we constructed to replace the fiberglass, the Kunzmobile rickshaw that we rigged up to replace the trailer, or the insane scaffold that we used in place of the crane, take note: these are the symbols of courage.

And the team that built them, and used them to erect the hab against all bets, can take pride in them, for these are the accoutrements of a company of heroes.

I believe that you shall be known and remembered, across vaster stretches of time and distance than any of us can imagine.

Because of what you have done, our movement will grow, and humans will go to Mars, and beyond.

Wherever they go, they will face similar challenges, and even greater risks. But they will prevail. Because they will carry on a tradition exemplified here.

This is just the beginning. As we stand here, at the launching point for the human reach to Mars, in a very real sense we are standing at the beginning of time.

The human race is young. Life is young. The universe is young. How so? The universe is 12 billion years old. Yet there will be luminous stars for another 120 trillion years. A person can live about 30,000 days. Compared to the future before it, the universe is in its third day. Life is in its first day, and the human race is in its first second.

Look around. It's the turn of the twenty-first century. Look back a hundred years, to 1900, see how much the world has changed. We live in cities with hundred-story-tall buildings, and fly across oceans at 500 miles per hour in ships weighing hundreds of tons. Compared to the world of 1900, we live in a science-fiction universe. So who can deny, seeing that, that if we dare, if we take this step out of the cradle, that a hundred years from now there can be a new branch of human civilization on Mars.

Now look back a thousand years, to the early medieval period, to the world lit only by fire. How magical our world would have to seem to someone from that time. Seeing the progress over that span, who can deny that if we open this new frontier, that a thousand years from now there can be thousands of new branches of human civilizations on thousands of new worlds orbiting stars over this region of the galaxy.

The human story has just begun. We started as a few tribes in east Africa, and spread from there, out of our natural tropical habitat, to become a global civilization. That was chapter one.

Now we will take the next step, to take on the challenges beyond and spread to new planets in space. That will be chapter two. In a very real sense, that chapter is beginning here.

And with that in mind, I want to take a few minutes to talk about the flag that now flies over this habitat—the red, green, and blue tricolor of Mars.

The idea for that flag was first suggested during our scouting expedition here last year. Some liked it for its association with Stan

Robinson's epic future trilogy of Martian history, *Red Mars, Green Mars, Blue Mars.* I liked it for that reason and others. The names of the books themselves refer to the future transformation of Mars, from its current state as a red, dead planet to one where the greenery of life has begun to spread, and ultimately to a fully living blue planet like the Earth. Red, green, and blue are also the primary components of light, and thus symbolize, light, enlightenment, and reason. The tricolor is also the most traditional form for the flags of republics, governments dedicated to the interests of the governed, and to the eighteenth century's noble ideals of liberty, equality, justice, and the rights of man—ideals still emphatically worth preserving and taking to Mars.

The flag does not depict the traditional Mars symbol, which represents the shield and spear of Ares, the war god, for *that* Mars is not what we have in mind. It also does not depict Mars as a planet. That is because it does not represent a physical object, it represents a people—a people of high ideals, a people of the future. It represents a people who will continue the work of creation by bringing life to Mars, and Mars to life, and in so doing, prove to all time the positive and precious nature of the human species and every member of it.

This hab is dedicated to them, a people who are yet to be, to those who will follow. I'm proud to have their flag flying for the first time over something we have built, an installation dedicated to their cause.

And I think it fitting, that we who are here today as witness to this birth, celebrate it, by firing the first salute to the Martian flag.

I then gestured to John Schutt, who was standing ready with a 12-gauge shotgun. A blast was fired into the air in the direction of the crater. At the sound of the report, everyone cheered.

The first salute has been fired. A new flag is flying high. Will sufficient force assemble to carry it forward to greatness? That is up to you.

Join us.

REFERENCES

1. Klein to Naugle, March 22, 1972, quoted in Edward Ezell and Linda Neuman Ezell, *On Mars: Exploration of the Red Planet 1958–1978,* The NASA History Series, Science and Technical Information Branch, NASA, Washington DC, 1984.
2. Vishniac to Soffen, March 1973, quoted in Ezell and Ezell, *On Mars.*
3. Norm Horowitz, et al., "Sterile Soil from Antarctica: Organic Analysis," *Science,* 164 (1969):1054.
4. Norm Horowitz, Roy Cameron, and Jerry Hubbard, "Microbiology of the Dry Valleys of Antarctica," *Science* 176 (April 21, 1972): 242–45.
5. Barry DiGregorio with Gilbert Levin and Patricia Ann Straat, "Mars: The Living Planet," Berkeley, CA: Frog, Ltd, 1997.
6. E. I. Friedmann, "Endolithic Microbial Life in Hot and Cold Deserts," *Origins of Life,* vol. 10, p. 223.
7. D. McKay, E. Gibson, K. Thomas-Keprta, H. Vali, C. Romanek, S. Clemett, X. Chiller, C. Maechling, and R. Zare, "Search for Past Life on Mars: Possible Relic Biogenic Activity in Martian Meteorite ALH84001," *Science* 273 (1996): 924–30.
8. E. Gibson, D. McKay, K. Thomas-Keprta, and C. Romanek, "The Case for Relic Life on Mars," *Scientific American* 277 (1997): 36–41.
9. Ben Weiss and Joseph Kirschvink, "Life from Space," *The Planetary Report,* November–December 2000.
10. C. Mileikowsky, F. Cucinotta, J. Wilson, B. Gladman, G. Horneck, L. Lindegren, J. Melosh, H. Rickman, M. Valtonen, and J. Zheng, "Natural Transfer of Viable Microbes in Space," *Icarus* 145 (July 2000): 391–427.
11. E. Imre Friedmann, Jacek Wierzchos, Carmen Ascaso, and Michael Winklhofer, "Chains of Magnetite Crystals in the Meteorite ALH84001: Evidence of Biological Origin," *Proceedings of the National Academy of Sciences* 98 (no. 5, February 27, 2001): 2176–81.
12. Lynn Margulis and Dorion Sagan, *Microcosmos: Four Billion Years of Evolution from Our Microbial Ancestors* (New York: Simon & Schuster 1986).

13. Chris McKay, "Oxygen and the Rapid Evolution of Life on Mars," in J. Chela-Flores and F. Raulin, eds., *Chemical Evolution: Physics of the Origin and Evolution of Life* (Netherlands: Kluwer Academic Publishers, 1996), 177–84.
14. Chris McKay, "Time for Intelligence on Other Planets," in L. R. Doyle, ed., *Circumstellar Habitable Zones* (Menlo Park, CA: Travis House Publications, 1996), pp. 405–19.
15. Stanley Miller, *The Origins of Life on Earth* (Englewood Cliffs, NJ: Prentice Hall, 1974).
16. Benton C. Clark, "The Viking Results: The Case for Man on Mars," in *The Future of the United States Space Program* 38 (AAS 78-156).
17. Robert Zubrin with Richard Wagner, *The Case for Mars: The Plan to Settle the Red Planet and Why We Must* (New York: The Free Press, 1996).
18. Marion Soubliere, *The Nunavut Handbook* (Iqaluit: Nortext Multimedia, 1998).
19. Pierre Berton, *The Arctic Grail* (New York: Penguin Books, 1988).
20. P. B. Robertson and R. A. F. Grieve, "Impact Structures in Canada: Their Recognition and Characteristics," *Journal of the Royal Astronomical Society of Canada* 69, (1975): 1–20.
21. P. B.Robertson and G. D. Mason, "Shatter Cones from Haughton Dome, Devon Island, Canada," *Nature* 255 (1975): 393–94.
22. T. Frisch and R. Thorsteinsson, "Haughton Astrobleme: A Mid-Cenozoic Impact Crater Devon Island, Canadian Arctic Archipelago," *Arctic* 31 (1978): 108–24.
23. W. Alvarez, *T. Rex and the Crater of Doom* (Princeton, NJ: Princeton University Press, 1997).
24. R. A. F. Grieve, "The Haughton Impact Structure: Summary and Results of the HISS Project," *Meteoritics and Planetary Sciences* 23 (1988): 249–54.
25. L. J. Hickey, K. R. Johnson, and M. R. Dawson, "The Stratigraphy, Sedimentology, and Fossils of the Haughton Formation: A Post-impact Crater-Fill, Devon Island, N.W.T., Canada," *Meteoritics and Planetary Sciences* 23 (1988): 221–31.
26. Pascal Lee, "From the Earth to Mars," *The Planetary Report* 22, no. 1 (January/February 2002).
27. Robert Zubrin, Steve Price, Ben Clark, and Roger Bourke, "A New MAP for Mars," *Aerospace America,* September 1993.
28. Robert Zubrin, *Entering Space: Creating a Spacefaring Civilization* (New York: Tarcher Putnam, 1999).
29. Louis Friedman, *Starsailing: Solar Sails and Interstellar Travel* (New York: John Wiley and Sons, 1988).
30. Bernard DeVoto, *The Year of Decision* (New York: Houghton Mifflin, 1950).
31. Francis Parkman, *The Oregon Trail* (New York: Viking Press, 1982).
32. Robert Zubrin, "Victory from Space," *Space News,* October 2001.
33. Hal Roey, SETI Institute, private communication, April 2002.

34. C. Cockell, et al., "Exobiological Protocol and Laboratory for the Human Exploration of Mars: Lessons from a Polar Impact Crater," *Journal of the British Interplanetary Society,* submitted for publication, January 2003.
35. C. S. Cockell, P. Lee, G. Osinski, G. Horneck, G. and P. Broady, "Impact-Induced Microbial Endolithic Habitats," *Meteoritics and Planetary Science* (in press), 2002.
36. Jack Stuster, *Bold Endeavors* (Annapolis, MD: Naval Institute Press, 1996).

INDEX

Page numbers in italics indicate illustrations.

ABOUT THE AUTHOR

Dr. Robert Zubrin is an internationally renowned astronautical engineer and the acclaimed author of *The Case for Mars,* which Arthur C. Clarke called "the most comprehensive account of the past and future of Mars that I have ever encountered." NASA has adopted many of the features of Zubrin's humans-to-Mars mission plan. A former senior engineer at Lockheed Martin, Zubrin is president of the Mars Society and founder of Pioneer Astronautics, a successful space technology research and development firm. Zubrin is the author of over one hundred technical and nontechnical papers in the areas of space exploration and nuclear engineering, and holds two U.S. patents. His other books include *Entering Space: Creating a Spacefaring Civilization* and *First Landing.* He lives with his family in Indian Hills, Colorado.

ABOUT THE MARS SOCIETY

The Mars Society is an international organization dedicated to furthering the exploration and settlement of Mars. Currently, the organization has more than 6,000 members and chapters in forty countries. Further information about the Mars Society can be found at its website at www.marssociety.org.